PROPAGATION
BASICS

PROPAGATION
BASICS

hamlyn

Steven Bradley

First published in Great Britain in 2002 by Hamlyn,
a division of Octopus Publishing Group Limited,
2–4 Heron Quays, London, E14 4JP

ISBN 0 600 60393 8

A catalogue record for this book is available from the
British Library

Printed in China

10 9 8 7 6 5 4 3 2 1

CONTENTS

INTRODUCTION

Plants are capable of reproducing themselves perfectly well without any interference or help from the gardener – too well, in the case of the plants we know as weeds – so the first question that anyone who is thinking of propagating plants from their own garden will ask is: 'Why bother?' The long answer is that it ensures the continuation of plants that might otherwise die out and provides a plentiful supply of material for keen gardeners to use or give away. The short answer is that it is great fun.

Propagation is something that even young children can try with perfect safety, and if you choose the right plants it can inspire in them an on-going interest in gardening. There are few things more dispiriting than watching and waiting for something to happen, but sunflowers, cress or even the cut-off tops of lunchtime carrots can all reward a child with fast results and a sense of achievement.

Once you have started to grow plants for yourself – whether you have a whole garden, a few containers on a patio or balcony, or a small windowbox – propagating is the next logical step. Why pay garden-centre prices when you can get the same results for little or no outlay? Anyway, watching young plants begin to root and grow gives an adult as much pleasure and satisfaction as it does a child.

It is impossible to be prescriptive about propagation because plants, gardens and seasons vary so much. Cuttings, for example, should be taken when they are most likely to root, which can vary by several weeks not only from garden to garden, as plants develop at different rates in the different conditions in which they are grown, but also in the same garden from year to year, as weather delays or hastens ripening. Although seasons and even months are recommended for different propagation methods, the timing indicated should be regarded as a guide only, and a good propagator will learn to work in 'plant time' rather than in calendar time.

This book aims take the mystique out of propagation. It describes the equipment and materials that are needed for success and explains clearly the different techniques, indicating throughout the different plants that can be propagated by each method to get the best results.

Above: Plants are capable of regeneration,
which means that it is possible
to propagate replacements of your
favourite plants.

1 GETTING STARTED

One of the great pleasures of gardening is having plants that were given as cuttings or seedlings by friends or relatives. These plants bring meaning and purpose to a garden, and walking and working there is like being among old friends as we are reminded of individual people and special occasions. Being able to propagate our own plants is a way in which we can repay these gifts and also provide new plants from old favourites to replenish and restock our borders.

Right: Many perennial plants can be propagated by division and will flower within a year.

Left: Numerous edible plants can be propagated easily from seed.

The key to successful propagation is to take it steadily. Do not try complicated procedures before you have mastered the straightforward techniques and do not be disheartened if something fails. Patience is important, both in waiting for plants to respond by growing new roots and leaves and in accepting that sometimes you just have to admit defeat and start again. It is impossible to take precautions against the unexpectedly and unseasonably warm day in spring or the fungal attack that suddenly develops and kills all your seedlings overnight.

If you are trying propagation for the first time, the most critical piece of advice you will need is to be realistic in terms of both your abilities and the equipment you already have and might need to acquire. You cannot possibly know which cuttings are difficult to root, because that knowledge comes only with experience, but you can check the back of the seed packet to see if there is a warning that experience is necessary to germinate a particular plant. Some plants, whether they are being grown from seeds or cuttings, need extra warmth before they will produce roots. This will mean investing in a heated propagation case, which will not be worth doing until you have decided that you want to propagate a wide range of plants on a regular basis.

If you have not tried propagating before, begin by germinating some seeds or taking some simple cuttings. As you gain experience, you will feel more confident and want to experiment with more complex and demanding procedures.

Many seeds, whether saved from your own garden or bought from a shop or seed merchant, will germinate without any special equipment. All you need are a clean tray or some pots, some new compost and the space to stand the trays until the seedlings can be planted out.

When you have successfully raised trays of seedlings and seen them grow to become mature plants in your garden, you will want to try other types of propagation, and taking simple cuttings will probably be the next step. Fuchsias and pelargoniums are ideal first subjects because they root so easily, but there are many other common garden plants that will propagate quickly and easily from cuttings, and before you know it you will have several types of cuttings rooting throughout the year, and you will be ready to turn your hand to division and layering. As you progress and gain experience you will find it increasingly easy to spot the shoots that will root quickly and form good plants. For most gardeners the question soon becomes not 'Should I propagate this?' but 'What am I going to do with all these plants?' It is a fascinating and rewarding aspect of gardening, coupled as it is with a real sense of achievement.

Left: Fast-growing plants like lettuce will grow from seed to maturity within two to three months.

Right: Roses are produced on a 'field scale' by specialist nurseries.

what is propagation?

In the controlled environment of a garden, where plants are more or less confined to particular areas, gardeners use propagation as a means of providing plants in both the short and long term. Bedding plants, annuals and many vegetables grow readily from seeds; they are planted out, enjoyed or eaten and then finally removed at the end of the season.

Perennials, which provide the longer-term structure of ornamental beds and borders, need dividing regularly to rejuvenate them and keep them growing strongly, and the extra plants generated in the process can be spread around the garden or exchanged with friends for new plants. When shrubs, which are included for the permanent framework they provide, outgrow their allotted space they can be replaced by new ones, grown from cuttings taken from the original plant or by layering.

NATURAL PROPAGATION

Many of the plants growing in the wild propagate themselves by setting seeds, which is dispersed by wind, birds or animals until it comes to rest in the first place that provides suitable conditions for germination. Seeds seldom fall directly to the ground, because growing too close to the parent plant would mean that they were both competing for the same water and nutrients, which might mean that neither eventually survived.

In the wild most plants set seeds once a year after flowering. Some plants are self-fertile – that is, they do not need pollen from a separate plant in order to do this – others, however, must be cross-pollinated with another plant of the same species before the seeds will set. Once they are ripe, the seeds are released by the plant so that they can find somewhere to germinate and grow.

Plants can also spread by means of underground stems, which send shoots up into the sunlight at regular intervals, and by means of shoots that arch over and root where they touch the ground. This type of propagation means that each new plant is absolutely identical to the parent.

SEEDS VERSUS VEGETATIVE PROPAGATION

The production of a seed depends on pollen fertilizing an egg cell, so every seed carries the genetic characteristics of both its parents. This means that with each generation slight

changes can occur, affecting everything from hardiness to height to flower colour. Companies that produce seeds carry out extensive research into the outcome of plant breeding, and, where they can guarantee that the resulting plants will be uniform, they sell the seeds as F1 (or first-generation plants from known parents).

More usually, however, the results are unpredictable, and the seedlings vary in size, shape and flower colour. Plant breeders find this phenomenon useful as a means of finding new colours and flower shapes. Huge quantities of rose seedlings, for example, are grown each year in the hope of coming across a single new cultivar that may prove marketable. Many plants readily set large quantities of seeds, and so this is a good method of producing lots of young plants quickly and easily. It is useful where the results do not all have to match or a large quantity of seedlings is required.

In contrast, vegetative propagation (the name given to all other means of reproducing plants) will produce young plants that are identical to their parents in every way, and these are known as clones. This includes plants grown from cuttings and by division and layering, and the techniques are useful for reproducing smaller numbers of identical plants.

Sometimes vegetative methods are the only way to propagate a plant. There are a few plants, such as *Acer griseum* (paper-bark maple), that set little fertile seeds, while others have problems with germination and may need intensive treatments, such as refrigeration, before they can be coaxed into growth.

TIMING

Although it is possible to propagate plants all year round, the time of year will often determine how well the propagation process works and how a particular plant responds. Most plants need to be actively growing in order to form roots and grow, but they will only respond at the right stage of growth. Conifers and some evergreens, such as *Pyracantha* spp., for example, are unlikely to grow successfully from cuttings unless the wood at the base of the cutting is at the semi-ripe stage. Other plants, which may be propagated by seeds, roots and hardwood cuttings, need to be dormant for a while before they can germinate or form roots and grow.

The most versatile group of plants are indoor or houseplants, which tend to grow throughout the year with no real resting period. As they have no true seasons (as outdoor plants do), they can be propagated at almost any time of the year.

Left: Over 100 years old, climbing rose 'Zéphirine Drouhin' is still propagated today.

Right: The only way to reproduce the contorted stems of *Corylus avellana* 'Contorta' is by vegetative propagation.

why propagate?

Nowadays, garden centres, nurseries and even do-it-yourself stores and supermarkets are full of beautifully growing and healthy-looking plants at almost any time of the year, and garden catalogues are filled with tempting illustrations of everything from well-grown trees and shrubs to plantlets that are ready to be bedded out.

So what, you might ask, is the point of propagating your own? There are several good reasons to propagate rather than buy.

COST

One of the main reasons for propagating from your own plants is the cost of even quite ordinary commercially raised plants. This is, of course, largely because you are paying for convenience. Buying a selection of container-grown plants gives you an instant garden at any time of the year, and if you are redesigning your garden or filling a brand-new one it is tempting to want to achieve an instant effect by buying well-grown plants. But

the 'instant garden' comes at a cost and gives no real sense of satisfaction. Although gardening is a hobby that can be enjoyed by anyone, not everyone has an unlimited budget; propagating your own plants is one way in which new plants can be produced each year for little cost.

Apart from the cost of compost and, perhaps, containers (although containers can be recycled and you will quickly accumulate almost as many as you will need), seeds require only a modest outlay and cuttings are usually free. You can plunder your own garden for material or ask friends if they would object to you taking some from their gardens. Often, if you do buy a plant, it can be used to provide cuttings or be divided to produce several smaller plants before you plant it. In this way you can get a number of new plants to go with the original and give a massed effect in the garden, rather than having a solitary plant that might not be very noticeable.

One of the great advantages to growing from seeds is that it is possible to produce hundreds of plants if you want. A single packet of seeds can contain anything from half a dozen to more than 200 seeds, which is ideal for bedding plants, and you can easily grow enough to give you a really colourful display through the summer.

ENSURING CONTINUITY

Many of the most popular and widely grown summer bedding plants are not reliably hardy, and if you want masses of colourful plants in your borders and containers you will need to grow half-hardy annuals from seeds each year. Hardy annuals and biennials are also grown from seeds, and, although many of these plants will self-seed, sowing seeds will allow you to impose a design or colour scheme on your borders.

In addition to the plants raised from seeds each year, some plants – notably fuchsias and pelargoniums – are not hardy, and taking cuttings each year is the best way of ensuring that you have new supplies of your favourite plants.

If you have a plant in the garden that is of sentimental value or that would be difficult to replace, propagating it is a good way of making certain that you do not lose it during a particularly cold season, and it will also enable you to create extra plants for elsewhere in the garden. These young plants are ideal for sharing and exchanging with friends and fellow gardeners, and it is in this way that forms and cultivars that are 'out of fashion' with the commercial nurseries are still passed around. Some cottage-garden favourites are still being exchanged in local sales long after they were regarded as commercially extinct by growers. This is how gardening heritage is preserved for future generations to enjoy, and without it countless valuable cultivars might have been lost for ever.

VARIEGATED AND COLOURED LEAVES

Plants with variegated leaves arise as result of a chance malformation in the cells, which would not be passed on in seed, so the only way to reproduce such plants is by vegetative propagation. The same is true of plants, such as *Corylus avellana* 'Contorta' (corkscrew hazel), that grow with contorted stems. Some plants with coloured leaves, such as *Fagus sylvatica* Atropurpurea Group (purple beech), will come true from seed, but the seedlings often contain a mixture of leaf colours – some green, some purple and some halfway between. The only way to ensure a complete batch of purple plants is to graft pieces of the original purple-leaved plant onto one-year-old seedlings of *Fagus sylvatica* (common beech).

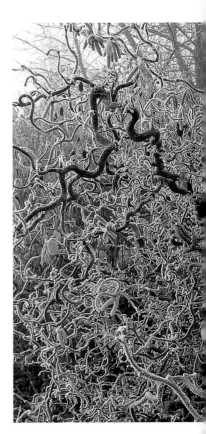

Sports

Plants occasionally undergo a genetic change that causes them to mutate and produce a flower, leaf or shoot that is different from the parent plant. Propagators try to take advantage of these mutated shoots to produce whole plants with the new characteristic, and this is the basis of many new cultivars. The most usual mutations, or sports, are variegated leaves on all-green plants or flowers that are a different colour. The climbing rose 'Martha', for example, which has pale pink flowers, is a sport of 'Zéphirine Drouhin', which has deep pink flowers.

Left: Honeysuckle is very easy to propagate from cuttings.

Right: Apple 'Worcester Pearmain' is propagated by budding or grafting because it will not grow from cuttings.

types of propagation

Plants can be propagated by seeds, by cuttings, by division, by layering and by grafting. Of these methods, only sowing seeds does not involve using a portion or even all of the parent plant. Only seeds, on the other hand, cannot always be guaranteed to come true to type.

SEEDS

Growing plants from seeds is the easiest way of producing new plants. If you buy a packet of seeds from a display in a supermarket or by mail order from a specialist catalogue you can be sure that the plants that eventually grow will look like the illustration on the packet. If you collect and sow seeds from plants in your own garden you cannot be certain that the resulting plants will be identical to the parent plants, unless they were species. However, the very uncertainty of

saving seeds from your own plants is part of the interest and enjoyment of the process.

Some plants will grow from seeds sown *in situ*; other types of seeds must be sown and germinated under cover and then planted out. The seeds of some plants can be sown over several weeks, giving a succession of plants, which is useful for crops such as lettuces and radishes.

CUTTINGS

Cuttings can be taken from almost every part of a plant except the flowers. The type of cuttings taken from the stems depends on the time of year, and it is possible to take basal, softwood, semi-ripe and hardwood cuttings. Many houseplants are propagated by leaf cuttings, which can be taken at almost any time of the year. Leaf cuttings are of three main

types: leaf petiole, leaf section and leaf slashing. Roots can also be used as cutting material.

DIVISION

Herbaceous perennials – that is, plants that live for two or more years – are most easily propagated by division, a process that involves lifting the entire plant from the ground and separating it into small pieces, each with a section of root attached. The technique not only provides several new plants but it also has the effect of rejuvenating the parent plant, which will have become overcrowded in the centre and which will, if not lifted, not grow well.

LAYERING

Layering is the only method in which the new plant is not removed from the parent until it has formed roots to support itself,

magnolias and some rhododendrons, and it is the best method for propagating houseplants such as *Ficus elastica* (rubber plant).

GRAFTING

This is a method of joining two different plants together so that they will continue to grow as a single plant. It is widely used for fruit trees, for example, as a means of determining the ultimate size of a tree by grafting a plant (scion) on to a rootstock that has only limited vigour, but rootstocks can also be selected because they are resistant to particular pests and diseases, will grow in specific soil types or are drought-tolerant. It is even possible to graft two or more

scions onto a single rootstock, as happens with 'family' apple trees, when several cultivars appear to grow on the same plant.

There are several types of grafting, including whip-and-tongue, apical wedge, side wedge, chip grafting, budding and shield budding, but it is a technique rarely undertaken by amateur gardeners. However, there are some plants, including some types of *Sorbus* (rowan) and fruit trees, that are difficult to propagate from cuttings, and grafting may be the only possible technique. The main difficulties are to ensure that the rootstock and chosen scion are compatible and that the various tissues of the two stems align properly to create a smooth union.

and, because there is no risk involved, layering is an ideal technique if you want to experiment with propagating. There are several methods of layering, some of which occur naturally, without any help from the gardener, when a stem bends down to touch the soil and produces roots. Eventually, a new plant is formed that grows independently of the parent, even though it may remain attached by the original stem. In the garden it is a simple matter to bend a flexible stem down to soil level and peg it in position until it roots. If it does not root, it can be released to continue growing and a different shoot pegged down.

Three or four plants can be propagated from the long, soft stems of climbing plants such as honeysuckle and jasmine by serpentine layering, and new plants of blackberries will root easily from tip layers. The technique can be adapted to propagate plants that have stems that are too stiff to bend easily down to the ground, such as

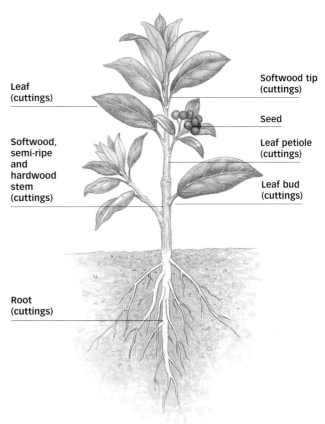

Leaf (cuttings)

Softwood, semi-ripe and hardwood stem (cuttings)

Root (cuttings)

Softwood tip (cuttings)

Seed

Leaf petiole (cuttings)

Leaf bud (cuttings)

PARTS OF THE PLANT USED FOR PROPAGATION

2 TOOLS AND EQUIPMENT

Most of the tools used for propagation are also used for general gardening, and you are likely to have many of them – secateurs, dibbers, rakes and hoes, for example – in your greenhouse or garden shed already. Until you have tried each of the main propagation methods and have decided that you wish to continue to practise one or more of them in future years, do not buy expensive items of equipment or tools.

Some of the items you will use for propagating, such as seed trays, pots and composts, will be useful for several different methods of propagation as well as around the garden in general, and there are other items that you probably take for granted, such as pencils or marker pens and labels, that are useful for recording what is propagated and when and labelling rows of seeds in the vegetable plot. There are, however, several items, such as a sharp knife for taking cuttings, sieves and presser boards, that are used only for propagating and that you may use only a few times each year.

Although sowing a few seeds or filling a few pots with cuttings will take up little room, and the seeds can be germinated and the plantlets grown on if the pots are stood on a windowsill, there is no doubt that having a greenhouse is convenient, and one that is heated is doubly so, providing an environment in which tender cuttings can be safely overwintered and autumn-sown seedlings protected until they can be planted outdoors.

For propagation outdoors you will need a number of other items, notably a rake, some garden line and a hoe. Again, they are not particularly specialized, but they are essential for getting the conditions right for successful propagation in garden soil.

Raking and firming garden soil before you insert cuttings or sow seeds will reduce the amount of settlement around them once they are in place. If you are sowing some of the smaller seeds you will have to rake the soil to a fine tilth (texture) before they can be sown or they will have trouble germinating.

Garden line – a fine, brightly coloured twine, held between two spools – allows you to work

Above: Most of the plants here, such as the pelargoniums, can be propagated from softwood cuttings taken when the plant is growing.

Left: Always use clean pots, as plants are very vulnerable to pests and diseases during propagation.

in a straight line between two points. Rows of cuttings and seedlings are easier to care for if they are grown in regularly spaced, straight rows, and the plants will usually grow more evenly because they are competing equally for light, water and nutrients.

It is inevitable that some weed control will be necessary to help cuttings and seedlings grow unhindered, and a hoe can be used between the plants to disturb the soil surface.

1 horticultural fleece
2 sieve
3 watering can
4 dibber
5 presser board
6 tamper
7 sharpening stone
8 polythene bags and sheeting
9 glass sheets
10 watering tray.

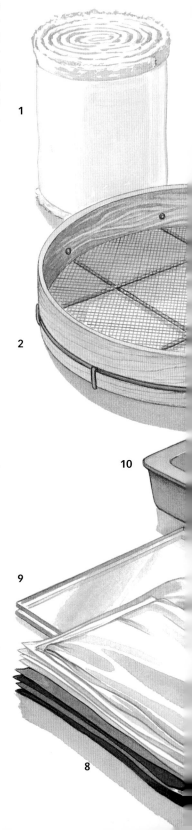

basic tool kit

Several of the larger, most useful items are discussed in detail on the following pages, but you will also find the items described here invaluable.

DIBBER
Although more commonly associated with transplanting seedlings, a dibber (a blunt-ended wooden or plastic implement) is useful for placing larger seeds in the compost with greater accuracy and to a specific depth, as well as for making holes in compost into which seedlings can be pricked out or transplanted.

GLASS SHEETS
A sheet of glass can be placed over a seed tray or pot to keep moisture trapped underneath so that the humidity level is high around the cuttings or seeds to encourage rapid growth. The glass will allow light to reach the compost surface, which is important for seeds that need light to germinate. Both cuttings and young seedlings must have plenty of light when they have started to grow.

HORTICULTURAL FLEECE
This lightweight material has two main uses: it will protect cuttings and seedlings from late-spring frosts (which can damage even the toughest of plants when they are young) and provide shade for the tender leaves of young plants, which can be scorched by strong sunlight.

POLYTHENE BAGS AND SHEETING
Polythene (plastic) bags have a wide range of uses in the garden, but they are particularly useful for gathering cuttings both inside and outdoors, as they reduce moisture loss and keep the material fresh until it can be dealt with. Held tightly in place with an elastic band over individual plant pots or seed trays, they are also useful substitutes for a propagation case.

Clear plastic sheeting can be stretched over wire hoops to protect plants, keeping them

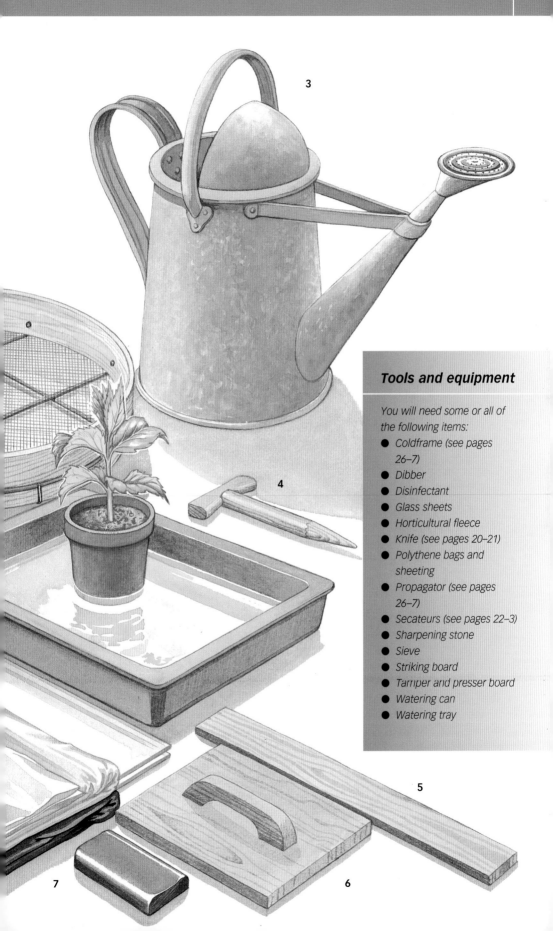

3

4

Tools and equipment

You will need some or all of the following items:
- Coldframe (see pages 26–7)
- Dibber
- Disinfectant
- Glass sheets
- Horticultural fleece
- Knife (see pages 20–21)
- Polythene bags and sheeting
- Propagator (see pages 26–7)
- Secateurs (see pages 22–3)
- Sharpening stone
- Sieve
- Striking board
- Tamper and presser board
- Watering can
- Watering tray

5

7

6

warmer and advancing their growth. Black plastic can be used as a mulch, with hardwood cuttings inserted through it into the soil below. This material is also useful for helping to warm up the soil in the spring, which is important for improved seed germination and rooting of cuttings outdoors.

SHARPENING STONE
If you have a good-quality knife and use it only for cuttings you will not often need to sharpen the blade, but it is useful to have one of these rectangular carborundum blocks to keep the blade sharp and clean (see also the next section).

SIEVE
A sieve is an essential part of the propagator's armoury if plants are grown from very small seeds, which need a fine surface on which to be sown and possibly a light covering afterwards. A mesh size of about 3mm (⅛in) should be adequate for all but the smallest seeds.

STRIKING BOARD
A straight, flat piece of wood (similar to a ruler), 3–5cm (1–2in) wide, is used to remove surplus compost from overfilled trays or pots when they are being prepared to receive either seeds or cuttings.

TAMPER AND PRESSER BOARD
Tampers and presser boards are used for gently firming the compost in a seed tray before sowing, because most seeds need a level seedbed to germinate and grow well. A presser board (a flat piece of wood with a handle on the back)

is used for square or rectangular seed trays, whereas a tamper (usually a circular piece of wood with a straight handle) tends to be smaller and is used for firming compost in pots.

Some presser boards (often called peg boards) have lots of small plastic or wooden pegs – mini-dibbers – on them. These leave a series of holes in the compost when it is firmed, into which seeds, seedlings or cuttings can be inserted, guaranteeing that the plants in the tray are evenly spaced.

A presser board is not as important for firming cuttings compost, because compost for rooting cuttings needs to be much looser and have much more air in it than compost used for seeds.

WATERING CAN
The rose on the end of the watering can is far more important than the watering can itself. Ideally, a fine rose should be fitted to the can to produce a fine spray of droplets for watering cuttings, seeds and seedlings effectively.

WATERING TRAYS
If fine seeds or small cuttings are placed in pots or trays it can be useful to have a larger container with water in the bottom. The propagation container can be placed in the tray so that the water will soak up through the compost without disturbing the seeds or cuttings. This soaking process can be repeated for several weeks until the plants are strong enough to be watered from above. However, soaking the compost in this way may wash out some of the fertilizer and slow plant growth.

knives

One of the most important – if not the most important – tools required for propagating plants is a knife. If you are buying a new one make sure you choose a model that is reasonably small and manageable, that fits snugly in your hand and that folds easily to drop back into a pocket when it is not in use.

There are special knives for specific purposes, such as grafting and budding, but to start with opt for a proper cuttings knife. The handle should be smooth and comfortable to hold, with no sharp corners or ridges or protruding rivets, which could make your hand sore or cause blisters. (If your hand is sore you will not grip the knife tightly and accidents are more likely to happen.) If this is your first garden knife, choose one with a straight, rather than a curved, blade because straight blades are easier to clean and sharpen.

The main reason a knife is preferred to secateurs for propagation in general and for taking cuttings in particular is the quality of the cut. A good, sharp knife will give a clean cut without ripping or tearing the plant tissue, and the softer the plant stem the more important the quality of the cut. Any crushing or bruising will cause rot where the cut was made, and this, in turn, will lead to a lower success rate, with more cuttings succumbing to basal rot and fewer actually forming roots.

TYPES OF KNIFE
Knives that are used for taking cuttings, especially softwood

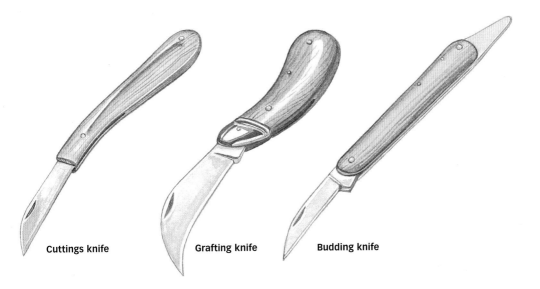

Cuttings knife   **Grafting knife**   **Budding knife**

cuttings, should have a straight blade that ends in a sharp point. The point is necessary so that the tip of the knife can get into small spaces to do such delicate jobs as removing leaves from cuttings without damaging the stem or other parts of the cutting in any way.

For grafting the ideal knife will be heavier than the one you use for taking cuttings, and it will have a broader blade and a thicker, stouter handle, because it is required to cut through harder, more mature wood.

Budding knives are usually the lightest knives used for propagation because they are used to cut through just a few layers of plant tissue, and the work involved in doing this is quite delicate. This type of knife often has a thin, curved blade and the end of the handle (opposite the blade) may be tapered or have a spatula end for lifting a flap of bark without straining the blade.

CUTTING CORRECTLY

No matter which type of knife you are using, it is important to feel comfortable with it and to hold and use it correctly to reduce the risk of cutting yourself. Your knife must always be sharp. If you use a blunt knife you will tend to use undue force, which can lead to mistakes and even cause injuries.

Placing the section of plant to be cut between the knife blade and your thumb and pulling upwards is the easiest way to cut, but the knife and your thumb must move in the same direction and at the same speed.

SHARPENING

A knife that is used regularly will need re-sharpening from time to time.

Apply a small amount of fine oil to the smooth surface of a sharpening stone, such as a carborundum stone. Use the tip of a finger to spread the oil over the surface of the stone and push the blade gently along the lubricated area of the stone. Repeat this several times on just one side of the blade. Turn over the blade and repeat the process on the other side. Finally, wipe the blade clean of any oil residue. If the blade is blunt, begin by sharpening it on the coarse side of the stone and finish the sharpening process on the smooth side.

Safety first

Whenever possible, cut towards yourself. This may seem to be at odds with common sense, but it gives maximum control over the knife and how it cuts, whereas cutting away from you (in the way you might sharpen a pencil) gives far less control of the cuts you make. If you do not feel confident about using a knife for the first few times, especially when you are handling soft, sappy plants, lay the plant onto a wooden board and cut down onto the board. The knife must be sharp to do this; otherwise the soft plant tissue will be crushed.

From left to right: Parrot-bill secateurs, anvil, blade and half-anvil and manaresi-type secateurs.

secateurs

Sharp secateurs (pruning shears) are an asset all round the garden. They can be used for everything from deadheading to light pruning, but they can also be used for just about all the cuttings you will ever take for propagation purposes as long as the blades are sharp. Most types will cope with shoots up to finger thickness, which is more than adequate for propagation purposes. Once you find secateurs that suit you, you will use them for years and they will be among the most useful and used tools.

CHOOSING SECATEURS

There is a wide range of makes and types of secateurs available,

and, although buying an expensive pair may seem extravagant, they will, with a little regular care and maintenance, far outlast a less expensive type. Before you buy, try them in your hand and choose a size that feels comfortable. Make sure that they are not too heavy and that the spring is not so stiff that you can hardly close the handles. Some manufacturers make right- and left-handed secateurs, and some types have an ambidextrous grip and thumb catch.

Most modern secateurs are made of light carbon steel, and one or both blades may have been given a non-stick coating to make cutting easier and to reduce the chance of sap

staining them. The hand grips may be fibreglass, moulded nylon or a lightweight aluminium alloy coated in moulded vinyl to make sure they do not slip while you are cutting. A shock-absorber between the handles will save you injuring your fingers if the blades close quickly.

SECATEURS FOR PROPAGATION

There are several types of secateurs:

- Parrot-bill secateurs have two curved blades, which pass one another closely, cutting like a pair of scissors.
- Anvil secateurs have a single, straight-edged cutting blade, which presses down on an

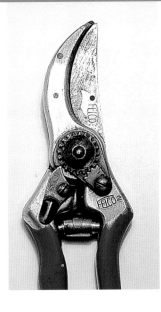

'anvil', which is made of a softer metal, such as brass.
- Blade and half-anvil secateurs have one curved cutting blade, which cuts past a second curved bar, which is fixed.

- Manaresi-type secateurs have two blades, both with straight cutting edges, which close together as they cut.

Some anvil secateurs have an in-built ratchet mechanism, which

makes it possible to cut through a branch in stages. This type of secateur is useful for someone who finds it difficult to squeeze the handles closed or who cuts a lot of woody plant material.

The parrot-bill type is ideal for propagation purposes because the scissor-like cutting action of the blades means that they do not crush the delicate stem tissue as they cut through it. Where a blade cuts by crushing the stem against an anvil there is inevitably some tissue damage. The straighter and cleaner the cut, the more quickly the cutting will heal and form roots.

Maintaining your secateurs

All types of secateurs will produce a good clean cut if they are used correctly and kept clean and sharp.
- *They should never be used for cutting wire (which will damage the blade) or for ripping open bags of compost (which may get into the mechanism).*
- *After you finish using them, wipe the blades to remove any remaining sap and allow them to dry before you close them.*
- *Keep the pivoting bolt oiled so that it moves easily and does not strain your wrist. If the secateurs have a spring, oil it at the start and end of the season to keep it rust-free and moving easily. When you use oil make sure that you remove any residue from the blades. Getting oil on your cuttings will stop them rooting.*
- *Once a year you will need to sharpen the cutting blade or blades, but if you find any really serious damage some manufacturers sell replacement blades, which you can fit yourself.*
- *Some of the leading manufacturers even have a servicing department so that you can send your secateurs away for a thorough overhaul.*

USING SECATEURS

The technique used for collecting plant material for propagation is different from that used for pruning. When you are pruning, the stem to be cut should be positioned close to the base of the blade where it can be held firmly. The blades are liable to be strained or forced apart if the cut is made with the tip. When you are taking the small stem that will be used for a cutting, using the tip will allow you to make delicate, accurate cuts.

Left: As seeds germinate and grow they will deteriorate if they are not transplanted to give them more space.

containers

Almost every gardener seems to collect containers, almost without even trying to do so. You start with one or two pots, but within a few years, as you buy new plants and pot up others, you find that you have dozens of pots of various sizes and shapes as well as numerous other containers. As long as they are thoroughly cleaned, all of them can be used for propagation purposes. The fewer cuttings you take or seeds you sow, the smaller the containers will need to be. Neither cuttings nor seedlings remain in their first container for long once they have rooted, so it makes sense to be economical with compost and to use as many of the smaller pots as you need. Even the tray-like containers that garden centres use to hold a number of pots can be re-used, especially for seeds that will germinate outdoors, because they tend to be wide and shallow and to be made of quite tough plastic. Put a piece of newspaper in the base in order to stop compost being washed out of the drainage holes.

PLANT POTS

There is a wide range of sizes and designs of plant pots, but the basic shape remains the same and has done so for many years. Those with sloping sides so that the top is wider than the base are ideal for allowing the plants room to grow and for removing them when necessary, although the narrower bases can make the pots unstable if the compost dries out. They are made of traditional clay (terracotta) or plastic.

Plastic pots are light, which is useful if they are standing on shelves or staging, and they retain moisture well. They are cheap to buy and will last for several seasons. They are convenient to stack and store

Choosing the right container

The container you choose to germinate seeds will depend on how many seeds you are sowing and their size. For just a few seeds or for large individual ones, pots are the obvious choice because they need less compost and take up less room. For a whole batch of the same plants a half- or full-size seed tray will be easier to use. Bear in mind that the more seeds you pack into one tray or pot, the sooner they [...] each other. Closely packed seedlings also suffer more quickly if there is a fungal infection than more widely spaced ones, where there is better air circulation around individual plants.

when they are not in use, and they are easy to clean, which is important when they are used for propagating. The disadvantages are that plastic pots do not look as attractive as clay pots, and they eventually become brittle and crack or break, especially if they are exposed to direct sunlight for long periods.

Clay has a pleasing, traditional appearance and is heavy enough, especially when wet, to give good stability. The pots are more expensive to buy than plastic ones, and water is lost through the sides by evaporation, which means that the plant inside will need watering more often. These pots are easily damaged by frost, and they chip and crack if they are accidentally knocked over or dropped. Clay pots are also more difficult to clean.

Whether you use clay or plastic pots is largely a matter of personal preference and what you have to hand. Both will be perfectly adequate for propagating seeds and cuttings, and you may find that the only thing you have to consider is the weight your propagation area will bear: a large number of clay pots filled with moist compost will be a considerable weight on light greenhouse staging.

SEED TRAYS

There are two main sizes of these rectangular trays, which are, as the name suggests, ideal for growing seeds or larger numbers of cuttings. The depth of compost is not as great as in a pot, but there is more room horizontally for the roots to spread. The smaller trays are half the size of the larger ones and are useful for small quantities of seeds or a few cuttings, but the larger trays are

Containers for propagating range from trays and half-trays to modular containers and plastic and biodegradable pots.

more flexible because they can be divided into smaller sections with purpose-made dividers or pieces of plastic cut to fit.

Both sizes are available in sturdy pre-formed plastic, which will last for several years, or in a much thinner, lightweight plastic, which is intended for one season's use only. Inserts are available to fit inside the larger seed trays and divide them into small cells for growing individual seeds or small groups for thinning later.

Sturdier versions of the cell packs are available, and these are strong enough to stand alone without the support of the seed tray. They are ideal for sowing single seeds or for rooting individual cuttings of a number of different plants.

BIODEGRADABLE POTS

These pots are generally made of compressed peat or wood fibre, and they can be bought in multi-packs ready for use. They are very useful for plants that resent root disturbance when they are repotted, because the whole pot can simply be left intact. The roots can penetrate the sides of the pot and grow into the new compost. Biodegradable pots are ideal for tomatoes and other vegetables that are started off indoors until the weather is warm enough for them to be moved outside.

The pots tend to dry out completely if they are underwatered, and they will go mouldy if they are given too much water, but as a short-term measure they work well.

Left: The top of this coldframe has been opened to get the pelargonium cuttings used to conditions outside.

propagators and artificial environments

The ideal propagation environment is one that reduces the stress on the young plants to an absolute minimum. This will vary according to the method of propagation being used, the plant and the conditions required, but providing an environment that is as beneficial as possible will improve your success rate. The more quickly young plants become self-supporting, the sooner they can be moved into a more natural environment to continue growing.

In the protected propagation environment the factors critical to plant growth – moisture, humidity and warmth – can be controlled and regulated.

- The amount of moisture lost from the plants and compost can be minimized.
- When moisture is held in the air around cuttings they tend to lose less from their leaves, so that more remains inside the cutting, keeping the cells healthy.
- Some seeds need a higher temperature in order to begin germination, and some cuttings root more quickly if there is a higher compost temperature in the 'rooting zone' (that is, the base of the cutting).

PROPAGATORS

Purpose-made propagation cases vary from the basic to the extremely sophisticated, and your choice will depend on your budget, the space you have available and your aspirations. If you intend to do a lot of propagating or plan to experiment with some of the more unusual and difficult plants, one of the more expensive units will be a good investment.

The most basic unit consists of a simple plastic lid that fits snugly over a seed tray. Look for one that has an opening vent in the top so that you can control the humidity inside. The lid can be removed once the plants inside are growing well and moved to cover another tray, while the young plants remain in the original one. Bigger versions have a large base tray and cover, and will hold several standard-sized seed trays.

Electric propagators have a base unit that contains thermostatically controlled heating cables and a plastic top. The seed trays fit inside the case and above the cables, which may need a layer of sand to cover them and distribute the heat more evenly. The heat provided by the unit must match the needs of the material being propagated, because too high a temperature will actually prevent rooting in some plants, such as conifers, rather than encouraging it.

COLDFRAMES

A coldframe is often the link between indoor and outdoor propagation. It is ideal for giving plants a little extra protection, and perhaps its main value is to acclimatize plants as they are moved from propagation indoors to growing out in the open. A coldframe is an unheated structure with solid sides and a glass or perspex top to let in the daylight.

The lights on the top are usually hinged at the back and can be opened to allow in varying amounts of air. Although larger coldframes can be permanent structures, smaller, portable ones are equally useful.

If the coldframe stands directly over soil it can be used for rooting hardwood cuttings

into the ground over winter, because, although they do not need protection, such cuttings will usually root more quickly if they are sheltered from extremely cold weather.

If the coldframe is positioned on a hard surface, trays of those seeds that will germinate outside without heat can be placed inside it, as can trays of young plants that are being hardened off. As the weather warms up, the top can be opened for progressively longer periods during the day to get the young plants used to conditions outside so that they are ready for the day when they are planted out, but if late frosts are forecast the lights can be closed to protect the developing plantlets.

CLOCHES AND POLYTHENE TUNNELS

Cloches or a length of polythene, stretched over low hoops set into the ground, will give you a useful outdoor propagation area for hardy subjects that do not need heat. Once the seeds have germinated, the cloche or polythene can be removed, leaving the seedlings to grow on *in situ* until they are large enough to be transplanted. If a cloche is positioned over the final growing area for young plants, it will help to warm the soil in spring to give them a head start.

GREENHOUSES

The most versatile structure a gardener can have, a greenhouse can be heated, covered or ventilated to suit whatever is growing inside. The only drawback is that the plants inside seldom all have the same requirements, so it is often easier to use a propagator for young

plants and to move them into the main greenhouse once they are growing strongly. To gain the maximum benefit from a greenhouse the glass should be cleaned regularly to allow in as much light as possible in the winter and the spring.

Mini-cloches

A mini-cloche for use indoors can be made by pulling a clear polythene bag over a pot or tray of cuttings. Alternatively, a clear plastic drinks bottle with the base removed but the cap left on (see above) can be placed over individual pots or, outdoors, a small patch of soil to create a warm microclimate for seeds or cuttings.

3 OTHER ESSENTIALS

In addition to the tools and equipment that you will need to carry out the full range of propagating techniques, seeds and cuttings also need compost and water to grow. Hormone rooting preparations are also useful for encouraging cuttings to produce roots.

Left: Greenhouses and other protective structures are ideal for extending the growing season.

Right: As plants grow and the roots expand, they will need to be transferred to larger pots.

Below right: Hormone rooting powders are helpful in speeding up the rooting process.

A visit to any large garden centre will reveal the huge range of proprietary composts that are available, and it is always worth using the appropriate compost for a particular job. It is possible to make your own compost for raising seedlings and cuttings, and if you are propagating large numbers of plants this will certainly be the most economical solution, although sterilizing garden soil to remove all pests and diseases can be a problem as the soil must be heated to a temperature of 82°C (180°F) for 30 minutes.

To begin with, however, you are more likely to choose a proprietary mix. Many gardeners prefer to avoid using peat because of the damage to the environment caused by the extraction. Composts containing peat substitutes, such as coir (coconut waste), bark, crop residues and, sometimes, municipal garden waste, are increasingly widely available, and it is also possible to find organic mixes. The proportions of the various ingredients in proprietary composts vary, and it is worth experimenting with different

mixes until you find one that suits your methods of working and the types of plants that you want to propagate.

It cannot be emphasized too often how important hygiene is, especially with seedlings, because they have no resistance to fungal diseases and can be wiped out quickly. The spores often overwinter in containers that have been used before or on staging in the greenhouse, so at the end of the season or whenever containers are emptied it is important to wash everything thoroughly in a solution of a proprietary garden disinfectant.

Left: Potting young plants with a good quality, well-balanced compost is essential if they are to continue growing strongly.

Right: Some plants, like this *Ceanothus impressus*, will not grow in standard composts because they need good drainage.

compost

Compost is more than the brown material in which plants grow. It also contains much more than soil, and if you visit a garden centre or large DIY store you will find several different proprietary formulations that have been tailor-made for different purposes.There are two main types of compost: loam-based and loam-free. The basic bulk constituent of loam-based composts is soil, which makes them heavier and better at retaining moisture and plant nutrients than loam-free composts, which are based on peat or a peat alternative, such as coir (coconut waste). Loam-free or soil-less composts tend to be light and easy to handle, but they can dry out quickly and lose nutrients. When loam-free compost does dry out, it can be difficult to re-wet properly without the help of a wetting agent or spreader.

SEED COMPOST

The basic constituents of a seed compost are peat and sand or loam, peat and sand, and this is all that is required for basic germination. However, as most seedlings spend at least ten days in the compost after germinating, they need some nutrients, so small amounts of ground lime and superphosphate are added when they are mixed. Superphosphate is a particularly important addition because it promotes the development of new roots.

CUTTINGS COMPOST

The essential requirements of a cuttings compost are that it should hold enough moisture to keep the base of the cutting moist and that it should have an open texture, so that there is plenty of air in the compost, both of which are necessary for the formation of a healthy ·root system.

The most common bulk constituents of cuttings composts are sphagnum peat (which is good for holding moisture) and sharp sand (a fine grade of lime-free grit). Although the proportions of peat and sand may vary from manufacturer to manufacturer, the proportions of each are usually more or less equal in common commercially available composts.

If you want to mix your own compost, however, you could change these proportions depending on the time of year. For example, you could use a 3:1 ratio of peat to sand in spring and summer, when the light

(Californian lilac) and *Fremontodendron* spp. (flannel bush), which require a compost containing about 75 per cent sharp sand if they are to root successfully.

OTHER ADDITIVES

As a substitute for sharp sand other materials may be added to a propagation compost to improve aeration and drainage. These include grit, perlite (a white, grit-like material) and vermiculite (a silver-grey form of mica, which can hold large volumes of water or air). Perlite and vermiculite are extremely light, which makes them useful additions to compost if several trays of seeds or cuttings have to stand on greenhouse staging.

FERTILIZER

Different composts contain different levels of fertilizer, according to their use and the needs of the plants they are designed to support. Potting composts for established plants, for example, contain a range of fertilizers to give the plants a balanced feed, but they are much too powerful for seeds, cuttings or young plants. Even multi-purpose composts, which are sold for use for all stages of plant growth, can be a little too strong for some young plants and too weak for established ones.

Proprietary seed and cutting composts will contain low levels of the nutrients that will be needed as soon as growth starts, making them ideal for propagation purposes. Once the young plants are fully established and growing well, they can be moved into stronger composts as their need for fertilizer changes and increases.

levels are high and the compost might dry out quickly, and equal proportions of peat to sand in autumn and winter, when drainage is important to prevent the cuttings from rotting.

SPECIAL REQUIREMENTS

Plants have different growing requirements, and these are sometimes quite specific and must be provided if germination or root formation is to take place. The large group of lime-hating plants – rhododendrons, azaleas and camellias, for example – must have a wholly lime-free (ericaceous) compost, or the seeds or cuttings will die.

Many alpine plant seeds are sown into a mixture of equal parts of loam and sharp sand so that the compost does not become too wet for germination, because these plants need really sharp drainage. Other plants that also need an open, free-draining compost because they are unable to tolerate too much moisture are *Ceanothus* spp.

Storing compost

The trouble with mixing your own compost is that you seldom use all of it at once, so it has to be stored until next time. A strong plastic bag, of the type in which proprietary composts are supplied, is ideal. Fold over the top and roll it down to the top of the compost inside and place a weight on top so that it cannot unroll. This should keep out air and infections, and the bag can be placed in a dry, frost-free place. Do not leave compost open to the air, as it is likely to be contaminated by fungal spores or bacteria, which will germinate as soon as you put the compost in a warm propagation environment.

hormone rooting preparations

Plants are able to react to their environment in various ways, and they are sensitive to external factors, such as light, heat and the movement of air. Their reactions to these stimuli are the result of chemical changes within the cells, with different circumstances triggering different chemicals that regulate reactions in the rest of the plant.

The chemicals that help a cutting to root are known as plant hormones, and they are found in concentrations at various points around the plant. The concentration is strongest in the growing tip at the end of the shoot, but the hormones are also found at the leaf nodes (joints) along the stem. Young shoots have greater concentrations than older ones, so these are the shoots to look for when you are taking cuttings.

THE ROOTING PROCESS

When you remove a cutting from its parent plant, as part of the plant's natural healing process it reacts by sending hormones to the site of the injury. The plant will start to heal the wound, and roots will form at the base of the cutting as a second phase of this process. The more hormones the cutting contains, the greater is the chance that it will root quickly and successfully. In order to improve the likelihood of this happening you should aim, wherever possible, to trim your cutting just below a node and leave the growing tip intact. The more hormones the plant contains, the more ways you can take cuttings; some will root only when the growing tip is intact, but others need only one or two nodes from along the stem, so that a single shoot can yield many cuttings – one tip cutting or several internodal ones, as they are called.

As the hormones collect at the cut surface, they stimulate

Left: Hormone rooting powders will encourage cuttings to form new roots more quickly.

Right: Pelargoniums root easily from softwood cuttings without the need for rooting hormones.

USING ROOTING PREPARATIONS

Applying the right amount of rooting preparation is a matter of practice. It is required only on the cut surface, so if you are using a liquid dip the end of the cutting in and allow any surplus to drip off. If you prefer to use powder tip a little into the lid of the container or another shallow dish. Again, dip only the end of the cutting into the powder and tap off the excess.

It is important to observe strict standards of hygiene when you are taking cuttings, so that there is no cross-contamination between batches. It is a good idea to decant a small amount of the liquid or powder into a small container for immediate use and to throw away what is not used. This is not as wasteful as it sounds if it saves you from losing a whole batch of cuttings to a fungal infection.

Some hormone rooting preparations contain a fungicide, which can be useful in protecting cuttings against rot.

the formation of root cells. These multiply to form a lumpy, whitish growth, known as callus tissue, which swells until individual roots become visible to the naked eye. As soon as they form, these roots begin the essential process of supplying the young plant with moisture and nutrients to feed it and start the next part of the process, which is to put on more leaf and shoot growth. The quicker the cutting can root and start to nourish the plant, the better the plant will grow, and a cutting that struggles to root will be weakened during the time it is without food and adequate levels of moisture.

Many plants have enough of these hormones to root successfully from cuttings, but some do not, and these benefit from the use of a rooting powder or liquid. Both contain chemicals that mimic the natural hormones and give the plant a boost to help it root more quickly.

Useful though rooting compounds can be, they should never be used on some plants, notably pelargoniums. Using rooting powder on a cutting from a pelargonium is actually counterproductive and will usually stop it from forming roots because it already contains high levels of its own rooting hormones.

THE ORIGIN OF ROOTS

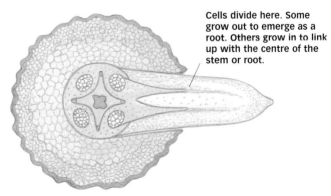

Cells divide here. Some grow out to emerge as a root. Others grow in to link up with the centre of the stem or root.

Hormone compounds are used to stimulate the response of the root within the plant.

Left: A watering can is ideal for watering containers. For small plants a rose should be fitted onto the spout.

Right: Plants like the African violet (*Saintpaulia*) have hairy leaves which are damaged by water droplets. They should be watered from below.

watering

When a plant is growing in its natural environment it can take the water it needs from the soil, and if the water is in short supply immediately around the plant it can send out new roots to find more. A plant that is growing in a container cannot do this, of course, and has to depend on the gardener to supply all its water. As every plant has different requirements, this is no easy task, and even experienced gardeners will admit (if they are honest) that they still get it wrong sometimes. Although with fully grown plants this is not usually anything more serious than being taken by surprise by an unexpectedly hot or windy day, when it comes to propagating too much or too little water can mean the difference between success and failure.

GETTING THE BALANCE RIGHT

Every plant needs water to survive. The cells in a plant are filled with water, and it is the means by which the sugars and starches on which the plant depends are transported around its system. A seed needs to absorb water before it can start to germinate, as the cells within it begin to divide and grow and the young plant begins to expand. If the supply of water is reduced the process of growth slows down, and if it dries up altogether growth stops. This does not mean instant death, because there is a brief period in which the seed could be revived if the water supply was restored. Growth would have been checked but would resume, although the plant might never be as healthy as it would otherwise have been.

Young seedlings are very sensitive to the amount of water they receive. Because they are growing quickly they need a regular supply, and a reduction will cause wilting within hours and the loss of the entire batch within a day or so. If the tissue at the base of a cutting dries out, the cutting will be lost, because without functioning cells at its base the supply of water from the roots to the leaves can never be established.

Too much water is as bad as too little, however, because the water reduces the amount of air around the seed or cutting, and without air it will die. Excess moisture will cause the plant's tissues to rot and decay irreversibly. Unfortunately, the instinctive reaction to noticing that a plant has been too dry is to give it plenty of water. What it really needs, however, is a gradual increase in the supply so that it does not drown.

WATER FOR ROOTS

Given a regular supply of water the cells in the plant's roots will divide and extend in search of more water and nutrients. These are absorbed and transported around the plant to feed the cells in the stem, shoots and leaves so that they can also expand. A shortage of water causes the roots gradually to shrivel and lose contact with the compost, so that, even if the compost is moistened, the roots cannot reach the water.

WATER FOR LEAVES

Plants' leaves are green because they contain a chemical, chlorophyll, which converts sunlight into sugars and starches, which the plant uses to grow. For this to happen, the cells need to be operating properly, and they cannot do that unless they have all the water they need. Full cells give the plant stability and hold the stem upright. A shortage of water causes the cells to begin to shrink and lose contact with each other – a process visible to the gardener as wilting. To a point this can be corrected by applying more water, but beyond the 'permanent wilting point' the plant will not recover.

hygiene

Hygiene is more critical in the early stages of a plant's life than at any other time because a seedling does not have the reserves of strength to fight off an attack by a pest or disease. All its energies are channelled into surviving and producing enough roots and new leaves to keep itself alive. Diseases and fungal spores are impossible to detect in the air around us, and small insect pests can get through any open window or door, so it is almost impossible to protect the plants from absolutely everything. By taking a few basic precautions, however, it is possible to reduce the problems they will face to a minimum, at least in the early stages of growth.

ENVIRONMENT

The propagation environment is an ideal breeding ground for diseases and fungal infections, which thrive in the warm, humid conditions. It is important to keep the whole propagation area as clean as possible, especially if it is going to be sealed for a while as the seedlings germinate or the cuttings root.

If you work in a greenhouse or other framed structure every winter you should give the whole inside a thorough scrub with a solution of a proprietary garden disinfectant to get rid of any insect eggs or fungal spores that might be hiding in the supports. Aerosol sprays or smoke canisters containing pesticides and fungicides can also be used if necessary. If you use a windowsill in the house make sure there is no mould around the edges of the window that could transfer to the compost.

TOOLS AND EQUIPMENT

The main tools any gardener uses are their hands, and these are just as likely to pass on infections as any other tools. If you have been handling infected material just before you start taking cuttings you will pass on the infection. Keep contaminated material well away from the intended cuttings and the area where they are to be prepared. Keep the prepared cuttings away from any other plant material.

The knife or secateurs you use to take the cuttings should be cleaned to remove any residual sap before you move on to another batch. Any other equipment you use during the procedure will also need to be cleaned thoroughly, including presser boards and dibbers.

POTS AND TRAYS

After each batch of seedlings or cuttings is moved on, soak the empty pots and trays in a cleaning solution and scrub thoroughly with a stiff-bristled brush to remove any remaining traces of compost, mould or old roots.

Allow them to drain and dry before storing them in a clean, dry part of the shed or garage. If they become dusty between use, rinsing them with clean water or dipping them in a dilute solution of disinfectant should be sufficient. Fungal spores can hide in the smallest places, so it is a good idea not to use any pots or trays that are cracked or split.

WORK BENCH

Fungal spores and bacteria are minute and can hide in cracks or crevices around the working area. For this reason, it is better to work on a completely smooth worktop, which can be wiped down with disinfectant before you start and scrubbed clean after you have finished. If the worktop is wooden, scrub along the grain with a stiff-bristled brush, working the disinfectant into any cracks to reach as many hiding places as you can.

4 SEEDS

Seeds can vary in size from microscopically small, dust-like particles to objects as large as a coconut, but all seeds are, in effect, a plant in kit form. No matter how small the seed, it contains everything that is required to produce an entire plant, from roots, shoots, leaves and flowers to, ultimately, more seeds.

Left: Poppies are very good at shedding seed. Over a period of a few years one plant can easily become a large colony. The seedheads are also very striking.

Below right: Many tree and shrub species, like the *Berberis thunbergii*, will come true from seed and can soon produce large numbers of young plants.

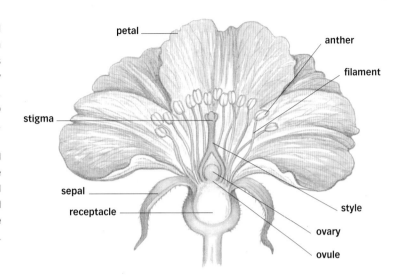

petal — anther — filament — stigma — sepal — receptacle — style — ovary — ovule

PARTS OF A FLOWER

A seed is formed when a pollen cell fertilizes an egg cell, and this is the sole objective of a plant when it flowers. If you remove the flowers from a plant as they fade and prevent it from setting seeds, it will try to produce more flowers, a procedure that is used by gardeners – when it is known as deadheading – to get a longer display. Once it has achieved its aim and set seeds to ensure the continuation of the species, the plant quite often stops flowering and diverts its energy into helping the seeds to develop. Some plants, such as some bromeliads and sempervivums, die after they have flowered and produced seeds.

When they are mature the seeds are shed from the parent plant to find a place to germinate and grow during the next season. Most seeds have a hardened outer casing to help them to survive until they find an environment in which they can grow, and some are also encased in a fleshy exterior cover, which has to rot away before the seeds can germinate.

Seeds are dispersed from the plant in many different ways. Some, like clematis and thistle, are borne by the wind on a light, fluffy tail or in a feathery ball, while others, like *Betula* (birch), are so tiny and light that they are blown on the slightest breeze. Some, like *Galium aparine* (goosegrass, cleavers), have seeds like sticky little balls, which cling fiercely to animals and people as they pass. Many of the fruits and seeds with fleshy coats rely on birds to eat the outer casing and drop the seed, either as they eat or once it has passed through their digestive systems. Members of the pea family, legumes, have pods that twist open as they dry and make an explosive snap, flinging the seeds as far as they can. Once dispersed, some seeds need to germinate straight away. *Acer pseudoplatanus* (sycamore), for example, has an entire embryo plant curled up inside each winged seed case. Other seeds can survive for several years before they germinate, especially if conditions are not favourable.

This is why we can buy packaged seeds each year: they are kept without light or oxygen in sealed packets. Inevitably, the seeds decay gradually if they are not sown because they are living organisms. The percentage that will grow decreases as time passes, even when they are kept in ideal conditions, so it is never worth storing your seeds for too long before you use them.

Left: Rhododendrons produce very small seeds and so are difficult to sow thinly.

Right: A large number of edible plants, like lettuce, can be grown from seed to maturity in one season.

germination and viability

The process of growth will begin when a seed has the right conditions to germinate, and these include air, light or dark (according to the species), water and the correct temperature. In this environment the seed will begin to absorb moisture from the compost and start to swell. Using the small reserve of food stored within the seed, cells begin to form the first root, which emerges to begin supplying more moisture to the new plant. Plants have the ability to sense gravity, so the root always grows downwards, and the shoot, when it emerges, will always grow upwards.

As a plant's root extends, it begins to grow fine feeder roots on each side to start absorbing nutrients from the compost. These replace the soon-depleted reserves from inside the seed and help to supply the new shoot that starts to grow. Soon, the plant will need other food supplies that only the leaves can supply, so it is important that it gets a pair of basic working leaves in action as quickly as possible. If the seed was planted too deeply, the seedling may not have sufficient reserves to keep it growing until it reaches the surface and it may wither and die. The first leaves (cotyledons) are quite simple in shape, usually rounded, and they serve only to get the rest of the plant into action. Once the leaves are functioning, the plant can begin to grow in earnest and produce its first pair of true leaves – the ones that it will always produce from now on. Light is important from this time, because even if

the seed had to be germinated in the dark the new plant needs light to live.

GERMINATION

Seeds will germinate whenever and wherever the conditions are right, and this includes a windowsill indoors. Even if you have an unheated greenhouse to which you can move the seedlings as they grow, you can give the plants an early start in spring by germinating them indoors. The seeds will need to be sown according to the directions on the pack and given light or dark conditions as needed, but the delicate new seedlings will need protection from intense heat, especially if the window receives full, hot midday sunshine. Laying a sheet of newspaper over the container will help reduce the effect of this heat and can be removed as the sun moves round. Once they are growing, the seedlings can be moved outside into a protected structure, such as a greenhouse or coldframe, but this must be done gradually to allow them to acclimatize to the colder conditions. If they are put out too quickly the shock will slow or stop the growth for a period, and the plant may even die if they are suddenly put into conditions that are very different from those in which they had been growing.

TREES AND SHRUBS

If you are growing the seeds of hardy trees or shrubs that need to germinate outdoors, you may choose to set aside a patch of the garden to do it in. These seeds have to have normal weather conditions, often including a cold winter, before they will grow, and some take

more than a year to germinate. If you have a separate area to sow them in you can make sure that they do not get dug up during routine cultivation before they have a chance to grow. It is difficult to generalize about the way you should deal with the seeds of trees and shrubs because even the size of the seeds can vary immensely. For example, the seeds of *Juglans regia* (common walnut) are quite heavy, with about 88 to a kilogram (40 per lb), whereas those of *Betula pendula* (silver birch) are light, with 5.2 million to a kilogram (2.5 million per lb).

DORMANCY

Seeds have an in-built mechanism, known as dormancy, that stops them germinating in unfavourable conditions. Breaking a seed's dormancy can be simply a matter of giving the seed enough water and darkness to grow, but some seeds require chilling or extra heat. The seeds of eucalyptus naturally germinate after bush fires in their native Australia, so they need heat before they will

grow. The seeds of some plants, including roses and daphnes, become dormant only when they reach full maturity, and this can be avoided by harvesting them before this stage has been reached and sowing the seeds straight away.

Germination

Seeds need the following if they are to germinate successfully.
- **Water:** seeds need water to make them swell and start the process of growing.
- **Warmth:** even the hardiest plants need a soil temperature of at least 7°C (45°F) to germinate, and others need a temperature as high as 21°C (70°F).
- **Light:** some seeds, such as lettuce and celery, will germinate only if they are exposed to daylight.
- **Darkness:** some seeds, such as Allium spp. (onion) and phlox, will germinate only if the light is excluded.
- **Air:** some oxygen must be present for germination to begin and continue.

saving seeds

After the first year of growing your own plants, providing it has been a warm enough season for the plants to have set their own seeds, there should be no need to buy more seeds until you want to change the types of plant you grow. The production of seeds depends largely on the weather, because a cold, wet season will mean both poor flower production and a lack of insects to do the pollinating. Such conditions are the exception, however, and in most years you will get enough seeds to save for next season – and usually some to spare and swap with your fellow gardeners.

COLLECTION

Collecting and saving seeds is not difficult, although the timing can be fairly critical, as a hot day in summer or autumn can often trigger the release of some of the more explosive dispersal methods, such as those of *Lathyrus odoratus* (sweet pea). The ideal conditions for collecting are found on a warm, dry day with little wind. Wet seeds can rot in storage and fall prey to fungal attack, so if you have no choice but to gather them in poor conditions, lay them out on a sheet of newspaper to dry before packing them away.

If possible, wait until the seeds are ripe before you begin to harvest them. It is tempting to gather them as they grow so that there is no danger of missing them, but if the seeds have not developed sufficiently to reach maturity they will not grow when you sow them.

It is easier to identify the ripeness of some seeds than of others. Members of the pea family, for example, produce pods, which hold the small round seeds, and these turn from green to a pale brown when they are ready. Rose hips and holly berries become brightly coloured. Poppies have seedheads like pepperpots, each with a row of small holes around the top through which the seeds can escape as the wind blows it from side to side, and this also changes colour to a

When the seedheads have formed, pick the flower stems of plants and enclose the heads in a brown paper bag, shaking them occasionally until all the seeds have been released.

Left: Seedheads of *Nigella damascena* should be collected before they burst open and shed their seeds.

Right: *Senecio doria* has seeds which are carried on the wind. If they are left on the plant for too long they will all blow away.

Saving seeds

The seeds of the following plants can be easily collected and kept for sowing the following year:
- Allium *spp. (ornamental onion)*
- Antirrhinum majus *(snapdragon)*
- Clarkia amoena *(godetia)*
- Consolida ajacis *(larkspur)*
- Eschscholzia californica *(California poppy)*
- Helianthus *spp. (sunflower)*
- Limnanthes douglasii *(poached-egg plant)*
- Nigella damascena *(love-in-a-mist)*
- Papaver *spp. (poppy)*
- Tropaeolum majus *cultivars (nasturtium)*

F1 hybrids

Seeds that are sold as F1 hybrids are guaranteed to give uniform results. Such seed is the result of the crossing of two true species or two stable forms of a species, and the resulting plants are vigorous and exhibit the desirable characteristics of the parents. F2 hybrids are a second-generation cross between two F1 hybrids. The resulting plants are not as vigorous as F1 hybrids, and they are usually more variable. Some F1 hybrids are not even capable of producing seeds.

pale brown, but by the time you have spotted the colour change, the seeds may be all gone.

One way to get round this is to fasten a paper bag over the seedhead as soon as you realize it is beginning to change colour. As it ripens naturally, any seeds that are shaken out will be caught inside the bag. Don't attempt to open the bag or you will let them all fall out. Instead, once you find the bag rattling with loose seeds when you shake it, cut the stem together with the bag intact and hang it upside down for the remaining seeds to fall out. Finally, untie the bag, lift out the empty seedhead and pour the seeds into a container for storage purposes.

Seeds that form in a cluster, such as those of marigolds, can be cut whole and laid on newspaper in a seed tray to dry completely and loosen away from the stem before being gathered up and stored. Seeds that are surrounded by a fleshy outer covering, like members of the rose family, need to be handled with care so they do not get bruised. Damage during harvesting can easily cause rot, which will affect the whole batch if it is not spotted in time.

STORAGE

Make sure you remove all other pieces of plant debris from among the seeds, and keep different plant types quite separate as you prepare them, or you may get a surprise when they eventually grow. Store the clean, dry seeds in clearly labelled, airtight containers in a cool, dark, frost-free place until they are needed. Smaller seeds can be stored in small, labelled envelopes within a larger, airtight container, such as an old empty biscuit tin.

The length of time the seeds will last in storage varies according to type and conditions, but some deterioration is inevitable over a period of time. Aim to use most seeds within two or three seasons to be sure you will get some germination, and be ruthless with the rest. You nearly always get more seeds than you will need, and they can be replaced the following year when the plants seed again.

Left: Vegatable plants like these are grown in individual cells. This makes them very easy to transplant.

sowing under glass

For your first attempts at growing seeds, choose a plant that needs no special propagation facilities and closely follow the sowing instructions on the packet. Use a compost recommended for seeds, because it will have the right combination of ingredients to give the young plants stability, good drainage and air.

PREPARING A SEED TRAY
Use a new or thoroughly scrubbed seed tray. Place it on a flat surface and heap compost into it. Hold both sides of the tray, lift and tap it onto the surface to get rid of any air pockets. Use a flat edge in a sawing motion across the top to get rid of the excess compost, leaving a level surface to sow the seeds on. If you have a flat presser board that fits inside the seed tray, place it on the compost and press down gently to firm the compost. This is not essential, and pressing the compost too hard can actually be counterproductive because all the air is pressed out, making it difficult for the young roots to penetrate.

The container you choose will depend on the amount of seed you are sowing and could range from a whole seed tray to a small pot. Whichever you use, remember to write the name of the plants on a label, because one seedling can look much like another and it may be difficult to tell them apart.

SOWING FINE SEEDS
Some seeds are so fine that they resemble dust in your hand – and the finer the seeds, the harder they are to handle and sow accurately. To make sowing easier, mix them with a small quantity of dry horticultural sand. As you sow the seed–sand mix in pinches between your fingers and thumb or from the crease in a piece of paper, you can make sure you are sprinkling it evenly over the whole tray because you

Sowing indoors

Even if you have a greenhouse, there may be times when you need to start some seeds growing indoors: the weather may be just too cold outside; the seeds are tender and you want to keep a close eye on them; you have no room anywhere else. Most houses have a windowsill wide enough to take a seed tray, but try to choose one that gets bright light, rather than full, direct sunshine, and remember to protect the surface beneath from any water damage. If you cannot avoid a window that faces the sun, keep a piece of newspaper handy so that you can cover the seedlings for a few hours during the hottest part of the day.

Sowing in a seed tray

1 Loosely fill a seed tray so that the compost is slightly above the rim of the container. Tap the container to remove air pockets. Use the edge of a presser board to strike off the excess compost so that it is level with the container and then lightly tamp

the compost with the presser board so that it is about 1cm (½in) below the rim of the container.

2 Sow the seeds across the surface of the compost. Fine seeds can be scattered from the crease in a piece of paper.

3 Sift compost over the surface of the tray so that the seeds are just covered.

can see where it lies. The depth at which you sow the seed is critical, because each seed contains only a limited reserve of food to sustain it as it sends out a shoot to reach the surface of the compost. Fine seeds will need little or no compost over them as they germinate, so follow the instructions on the pack. Small seeds that are sown too deeply will run out of food before they reach the surface, and although it may appear that no germination has taken place there might, in fact, have been 100 per cent germination, but none of the seedlings ever saw daylight.

In general, the smaller the seeds are, the less covering they need, and some actually need no compost over them at all. They may, however, still need darkness in order to germinate, and this can be provided by placing a sheet of newspaper over the pot or tray to block out the light. If they do need a light covering, sieve compost carefully over the top so that the covering is light and even, with no lumps.

SOWING LARGE SEEDS

Large seeds are easier to see and handle, and they can be placed accurately into the seed tray without much difficulty. To get the spacing right, lay a row along the long edge of the seed tray and a row down one of the short sides. Follow these rows to keep the spacing correct as you fill the tray. Cover the seeds with a layer of compost and press the top lightly with the flat of your hand or a presser board to firm it gently.

WATERING

The last thing you want to do after carefully sowing your seeds is to splash them back out by watering the tray too vigorously. To avoid doing this, lay the tray down on the floor outside and fit a rose to the watering can. Tip the can and start the water flowing to one side of the tray before moving it across to cover the tray and away to the other side before you stop the flow. This way, the water is already leaving the can in a steady flow as it reaches the compost, and is less likely to give a spurt that could wash out the compost.

Alternatively, if you have a shallow, flat container that is larger than the seed tray, stand the tray in it and slowly pour water into the larger container. Leave the seed tray for about 15 minutes so that the compost in it soaks up as much water as it needs, lift it out and let it drain, leaving the compost in the tray thoroughly moist.

SOWING IN POTS

The principle is exactly the same as for seed trays, but on a smaller scale. Pots are ideal for small quantities of seeds or for sowing larger seeds individually.

Poor germination

There are several reasons for poor, non-existent or patchy germination:

- Old seeds will not give good or consistent germination, because the percentage that are viable (able to grow) falls year by year.
- Poor storage means that even seeds that are not old may have dried out and will not germinate.
- Erratic watering – with compost alternately soaking wet and dry – will cause the seedlings to drown or shrivel.
- Draughts kill young seedlings by chilling them, and the effect will be worse in the area of the tray that is most affected.
- Incorrect light levels affect germination, because every seed has a definite requirement, some needing more, some less. Read the back of the packet carefully if it is a plant you have not grown before and follow the instructions. You cannot just sow the seeds and cover them: some seeds need little or no covering whereas others need to be completely covered.
- Incorrect or fluctuating temperatures affect germination because each seed has a range it prefers and another it can tolerate. Too hot or too cold, and it will not grow.

They can be grouped together to save space in a seed tray with no drainage holes.

Seeds that hate root disturbance can be sown into individual pots, or a multi-cell seed tray so that the roots are kept as a complete block when the seedling is planted out.

Some plants can be propagated by seeds all year round because they are never fully dormant and grow, albeit at varying speeds, for 12 months of the year. Many of the plants we grow indoors as houseplants can be propagated at almost any time, although there are certain times of the year when they are easier to propagate than others.

Other plants can be more complicated because their growth patterns are closely linked to soil and air temperature and the length of the days in each season. Even some quite hardy plants benefit from a helping hand when they are being propagated, and these should be sown under protection and nursed along until they are strong enough to survive outdoors in the garden.

STORING SEEDS

If you don't want to sow all the seeds in a packet at the same time, carefully close it again once you have enough, fold over the top and store it in an airtight container, in a dark, frost-free place.

The seeds should keep for at least a year. There is no rule that you have to sow the entire packet of seeds if you open it, and many will last for several seasons as long as they are stored in good conditions. This is often the case with vegetable seeds, such as tomatoes and peppers, of which you will need only a limited number of plants to satisfy your needs.

HOUSEPLANTS

The following houseplants are those that are most often grown from seeds, and they should be sown at the temperature indicated. Other popular houseplants, such as *Saintpaulia* (African violet) and *Streptocarpus*, are usually propagated from cuttings (see pages 84–5).

Plant	Temperature
Asparagus densiflorus Sprengeri Group	16°C (61°F)
Asparagus setaceus (asparagus fern)	16°C (61°F)
Begonia pendula cultivars	21°C (70°F)
Campanula isophylla (Italian bellflower)	Cool greenhouse
Capsicum annuum (Christmas cherry)	21°C (70°F)
Cyclamen persicum	12–15°C (54–59°F)
Hypoestes phyllostachya (polka-dot plant)	15–18°C (59–64°F)
Impatiens walleriana cultivars. (busy lizzie)	16–18°C (61–64°F)
Pericallis hybrida (cineraria)	13–18°C (55–64°F)
Solenostemon scutellarioides (coleus)	22–24°C (72–75°F)

BEDDING PLANTS

Sowing seeds of annuals under glass gives the plants a head start over those that are sown direct in the ground. Transplant the following to their positions outdoors when frosts are no longer likely.

Plant	Sowing time	Temperature
Ageratum houstonianum cultivars (floss flower)	Late winter to mid-spring	18–21°C (64–70°F)
Alyssum spp.	Early to mid-spring	15–18°C (59–64°F)
Antirrhinum majus cultivars (snapdragon)	Late winter to early spring	15–18°C (59–64°F)
Bassia scoparia (burning bush)	Early to mid-spring	15–18°C (59–64°F)
Begonia semperflorens	Early spring	21°C (70°F)
Cosmos spp.	Early to late spring	18°C (64°F)
Dorotheanthus bellidiformis (Livingstone daisy)	Late winter to early spring	18°C (64°F)
Lobelia cultivars	Late winter to early spring	15–18°C (59–64°F)
Nemesia spp.	Early to mid-spring	18°C (64°F)
Nicotiana spp. (tobacco plant)	Mid-spring	18°C (64°F)
Pelargonium spp.	Late winter to early spring	13–18°C (55–64°F)
Petunia cultivars	Early to mid-spring	15–18°C (59–64°F)
Salvia splendens cultivars	Mid-spring	15–18°C (59–64°F)
Senecio cineraria	Late winter	18°C (64°F)
Tagetes erecta cultivars (African marigold)	Late winter	18°C (64°F)
Tagetes patula cultivars (French marigold)	Mid-spring	18°C (64°F)
Verbena cultivars	Autumn or early spring	18–21°C (64–70°F)

VEGETABLES

Sow seeds of the following vegetables under glass to ensure that you have plants with the longest possible growing season. They can be moved outdoors when no more frosts are likely.

Plant	Sowing time	Temperature
Aubergine	Mid- to late spring	18°C (64°F)
Celery	Early to mid-spring	18°C (64°F)
Courgette	Mid- to late spring	18°C (64°F)
Cucumber	Mid- to late spring	18°C (64°F)
Melons	Mid- to late spring	18°C (64°F)
Pepper	Mid-spring	18°C (64°F)
Sweetcorn	Mid- to late spring	15°C (59°F)
Tomato	Early to mid-spring	18°C (64°F)

Scarifying

Some seeds develop a hard outer casing, which stop water from entering. Sometimes, immersing the seed in warm water for about 12 hours (overnight) will be enough, but particularly resistant varieties need extra help. Scarification involves cutting (nicking or chipping) just barely through the outer casing carefully with a knife or rubbing the seeds with coarse sandpaper.

Left: When pricking out seedlings always loosen the roots first and never handle the seedlings by their stems.

pricking out

As seedlings develop and grow they begin to crowd each other in their containers. This has two significant results:

- The lack of light forces the seedlings upwards, so that they become tall, weak and leggy;
- Insufficient supplies of food and water slow the growth rate.

Both factors leave the young plants vulnerable to attack by pests or diseases, which they might otherwise be able to fend off, and to prevent this the plants are normally moved into new containers to give them more room. This procedure, known as pricking out, may involve moving the seedlings into individual pots, another seed tray or a cell-pack (a seed tray-sized unit that is made up of small, individual compartments).

TIMING THE MOVE

Each time a plant is uprooted and moved into a new container it suffers a transplanting check (a hiccup in its smooth rate of growth) as it adapts to and roots into its new surroundings. Your aim should be to reduce this check to an absolute minimum by timing the move as well as you can and by causing the least possible damage to the seedling.

When you move seedlings a good rough guide is to wait until the second set of leaves are just starting to open. The first leaves that emerge are known as the seed leaves, and they bear little or no resemblance to the second or 'true' leaves of the plant. The seed leaves are often plain and rounded, and their main function is to start supplying food to the plant so it can continue growing.

By the time the first true leaves have developed the root system will be sufficiently established to support the seedling but not so large that it becomes difficult to remove it from the container.

HANDLING THE PLANTS

At this stage the seedlings are delicate, and any rough handling will kill them. It is important to loosen them from the original container before you lift them up, or you may tear the roots. Use a large plastic plant label or a dibber (see page 18) to lever under the roots so that you can lift them slightly.

Gently grip the seedling by one of its seed leaves and tease it out of the compost. Never grip it by the stem. If you bruise or break the stem, you will lose the plant; if you accidentally damage a seed leaf, not only does it have another, but once the true leaves appear they are no longer needed anyway.

REPLANTING

Before you begin, water the seedlings and prepare the new container, filling it with more fertile compost. Working on one seedling at a time, lift the seedlings to their new home. Make a hole in the new compost that is large enough to take all the roots of the plant, lower it into the hole and gently press some compost from one side to hold the seedling in place. Make sure that there is no air cavity left at the bottom of the hole, or the seedling will dry out and struggle to establish. On the other hand, don't press it too firmly. It is better to settle the plants into the compost by watering the whole tray once it has been completely

planted. Use a fine rose on the watering can to give the plants a gentle shower rather than a torrent of water, which might wash them straight back out of the compost.

SPACE-SOWING

If you are sowing seeds such as vegetables directly outdoors into a prepared seedbed, it may be easier to space- or station-sow the seed.

This involves sowing small pinches of seeds at intervals along the row, rather than in a continuous line. These will germinate in clusters which can be thinned to allow the strongest from each group to survive. If you space the clusters according to the final recommended spacing for the crop, the remaining seedlings can be

allowed to grow to maturity without the need for any further thinning or replanting.

This way, there will be far fewer wasted seedlings to be thrown away, the remaining plant will not have had a growth check and the gaps between plants will be perfect.

Handling seedlings

So that you do not damage seedlings when you handle them:
- *Loosen the roots first*
- *Hold the seedling by the seed leaves*
- *Tease out the roots gently – don't pull them*

Pricking out

As soon as the seeds are large enough to handle without damaging them, they can be transferred into containers of a more fertile compost.

1 *Tap the sides of the tray to loosen the compost slightly so it is less consolidated. Use a small dibber or plant label to loosen the compost further by gently digging under the roots of each seedling. Ease the seedling upwards out of the compost, holding it by one of its seed leaves.*

2 *Make a hole in a tray of fresh compost and lift the seedling (still holding it by the seed leaf) so you can lower its roots into the hole. Once all the roots are in, gently firm the compost around them using the dibber. Space the seedlings evenly and water them using a fine rose. This will help settle the compost gently.*

Left: Ventilation is essential to toughen up seedlings before placing them outdoors.

Right: Cloches are ideal to give plants an early start in spring.

which to germinate, almost as soon as they are growing the requirements change. Instead of warmth, the seedlings' main need is for light, and they can be moved to a slightly cooler but well-lit position. Keeping them too warm encourages soft, sappy and often weak growth. Cooler conditions cause the cell walls to thicken slightly, resulting in a shorter, broader plant, which has a greater ability to resist disease.

Once you have pricked out the seedlings into their second containers, they can be kept in the cooler position, perhaps in a greenhouse or coldframe. You must keep an eye on weather forecasts, however, because a rapid drop in temperature aware of cold or frosty nights will kill the seedlings. Keep some extra protection handy – a greenhouse heater, horticultural fleece or sheets of newspaper, for example – so that the plants can be protected from sudden frost.

hardening off

A young plant that has been growing in a warm, protected environment is ill equipped to face the conditions in the garden, and putting it straight outside would be almost certainly kill it. The strain to its system of having to deal with colder conditions, more wind, bright sunlight and possibly erratic watering often proves too much.

The procedure of avoiding this sudden change in conditions is known as hardening off, which involves acclimatizing the plant by gradually introducing it to outdoor conditions so that it is strong enough to survive.

STARTING OFF
Although seeds tend to need a warm, moist environment in

VENTILATION
As the seedlings grow and the weather outside gets warmer,

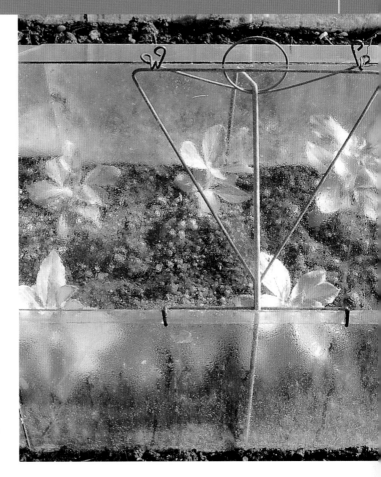

the ventilation around the plantlets can be increased each day so that they experience conditions more similar to those in the garden. Eventually, the containers can be moved outside to a sheltered spot during the day to make the most of the warmer weather, although they will need to be moved back under cover at night in case there is a frost.

ENCOURAGING STRONG GROWTH

Left to their own devices, some plants will grow to have a single, tall, weak stem. It is possible to 'stop' the plant by removing the growing tip, which has the effect of encouraging all the dormant side buds on the stem to start growing and producing branches, producing a stocky, multi-branched plant.

The same principle can be extended to the root systems of seedlings of some trees and shrubs, which have a tendency to produce a long, single taproot with few, small, side roots. Plants like this often take time to become established when they are transplanted, but with a little preparation beforehand the problem can be overcome. A few weeks before the seedlings are

due to be moved, they can be eased up: push a fork into the ground near the base of the seedlings and lift them slightly. This will break the taproot and encourage the side roots to start growing so that a fibrous root system develops, which can adapt to transplanting much better than a taproot.

Root pruning pots

There is a type of pot that has a small pocket or cavity at the base. This forms a gap between the compost and the propagation bench or sand bed on which the pots stand. The seedling's taproot will grow straight down to the bottom of the pot, out of the compost and into the air cavity, where the root tip will wither and die from lack of moisture. This forces the side roots to develop, giving a much more fibrous root system, which benefits growth rather than harming the plant.

BRUSHING

When young plants are grown under protection, especially when they are at the seedling stage, they sometimes become tall and leggy simply because they are so protected from the elements. In nature, seedlings would grow more slowly because of the constant disturbance they would suffer from the wind. The young plants' natural reaction to the constant movement is a thickening of the cells, which makes them shorter and stockier.

It has been found that if you brush your hand lightly over seedlings once or twice each day you will simulate the action of the wind and thereby encourage the stems to thicken and the plants to grow more strongly, as they would do in nature.

Left: Sowing seeds in drills makes the seedlings easier to see when they germinate, as well as making it easier to control any weeds.

sowing outdoors

Some plants will germinate quite easily, provided the conditions are right, when they are sown directly into the soil outdoors. They can either be sown directly into their flowering position or they can be grown in a seedbed for transplanting later when they are large enough. Tree and shrub seeds rarely need any heat to start into growth – indeed, they often need a period of cold (such as a winter) before they will grow. The main constraint is the soil temperature, which must reach a certain point for each plant for germination to start.

Hardy annuals can be sown into the border soil in spring and will provide an attractive display of flowers throughout the summer. They will often self-seed and emerge in the same area of the garden for several years to come. To get the most effective display from hardy annuals, they are best grown in large groups or drifts, with plants of different colours and heights being arranged to complement or contrast one another.

PREPARING A SEEDBED

The finer the seeds you are sowing, the finer the surface of the soil will need to be. If the surface is too coarse, the seeds may find themselves in a dry air pocket, where they will struggle to survive. When you have prepared the area by digging it, rake the surface with a fine-toothed rake to break down any remaining lumps and to make the surface level and even.

Seed compost

Loam- or soil-based composts are better for seeds that will be placed outdoors to germinate because the fertilizers they contain are less likely to be washed out through the winter months.

HARDY ANNUALS AND BIENNIALS

The following can be sown outdoors as soon as all danger of late frost has passed. Annuals will produce flowers the same year; biennials, such as *Digitalis* spp. (foxglove) and *Myosotis* spp. (forget-me-not), will produce leaves in the first year but will not produce flowers and set seed until the second year.

Plant	Sowing time	Soil temperature
Amaranthus caudatus (love-lies-bleeding)	Mid-spring	7°C (45°F)
Bellis perennis (daisy)	Late spring to early summer	8°C (46°F)
Bracteantha bracteata (strawflower)	Mid- to late spring	7°C (45°F)
Calendula officinalis (pot marigold)	Early spring	5°C (41°F)
Clarkia amoena (godetia)	Early to mid-spring	7°C (45°F)
Clarkia unguiculata	Early spring	5°C (41°F)
Consolida ajacis (larkspur)	Early to mid-spring	7°C (45°F)
Dianthus barbatus (sweet william)	Late spring to early summer	10°C (50°F)
Digitalis purpurea (foxglove)	Early to midsummer	10°C (50°F)
Dipsacus fullonum (teasel)	Early to midsummer	10°C (50°F)
Erysimum x allionii (Siberian wallflower)	Late spring to midsummer	8°C (46°F)
Erysimum cheiri (wallflower)	Late spring to early summer	8°C (46°F)
Eschscholzia californica (California poppy)	Early spring	5°C (41°F)
Gypsophila elegans	Early spring	5°C (41°F)
Helianthus annuus (sunflower)	Mid-spring	7°C (45°F)
Iberis sempervirens (candytuft)	Early to late spring	7°C (45°F)
Lavatera trimestris (mallow)	Early to mid-spring	7°C (45°F)
Limnanthes douglasii (poached-egg plant)	Early spring	5°C (41°F)
Lunaria annua (honesty)	Early to midsummer	10°C (50°F)
Malcomia maritima (Virginian stock)	Early spring to midsummer	7°C (45°F)
Matthiola incana (stock)	Mid- to late spring	7°C (45°F)
Myosotis sylvatica (forget-me-not)	Late spring to midsummer	8°C (46°F)
Nigella damascena (love-in-a-mist)	Early spring	5°C (41°F)
Oenothera biennis (evening primrose)	Early to midsummer	10°C (50°F)
Papaver nudicaule (Icelandic poppy)	Late spring to early summer	7°C (45°F)
Primula cultivars (polyanthus)	Late spring to midsummer	8°C (46°F)
Tropaeolum majus (nasturtium)	Mid-spring	7°C (45°F)
Verbascum spp. (mullein)	Mid-spring	7°C (45°F)
Viola x wittrockiana (pansy)	Early to midsummer	10°C (50°F)

SOWING IN DRILLS

A drill is a narrow channel drawn in the soil using the end of a bamboo cane or the point of a V-shaped hoe. The seeds are sown evenly along the channel by sprinkling them between the fingers and thumb (for finer seeds) or placing them at regular intervals (for larger seeds).

Once the seeds have been sown, they need to be covered, and this can be done quickly and easily by turning over the rake and using the flat side in short strokes to pull the displaced soil back down over the drill. Where you are sowing groups of several different types side by side, mark out the individual patches with

sand and draw the drills at angles to each other so the lines will become less visible as the plants grow.

BROADCAST SOWING

Seeds can be sown by being scattered evenly over an area, rather than in straight lines, and this method, which is known as

VEGETABLES

To get the longest possible growing season, sow vegetable seeds as soon as the soil temperature is warm enough according to the chart below.

Plant	Sowing time	Soil temperature
Asparagus peas	Mid- to late spring	8°C (46°F)
Beans, broad	Late autumn or mid- to late spring	8°C (46°F)
Beans, dwarf	Mid-spring to midsummer	8°C (46°F)
Beans, runner	Late spring to early summer	8°C (46°F)
Beetroot	Early spring to midsummer	7°C (45°F)
Calabrese	Mid-spring to midsummer	8°C (46°F)
Carrot, early	Early to late spring	7°C (45°F)
Carrot, maincrop	Mid- to late spring	7°C (45°F)
Chard	Mid-spring or late summer	8°C (46°F)
Chinese cabbage	Early to late summer	10°C (50°F)
Kohlrabi	Mid-spring to midsummer	8°C (46°F)
Lettuce	Early spring to midsummer	7°C (45°F)
Mangetout	Mid-spring to midsummer	8°C (46°F)
Onion, autumn-sown	Late summer to early autumn	10°C (50°F)
Onion, bulb	Late winter to early spring	5°C (41°F)
Onion, salad	Late winter to early summer	5°C (41°F)
Parsnip	Mid- to late spring	8°C (46°F)
Peas, early	Early to mid-spring	7°C (45°F)
Peas, maincrop	Mid-spring to midsummer	8°C (46°F)
Radish	Early spring to early autumn	7°C (45°F)
Spinach, summer	Early to late spring	7°C (45°F)
Spinach, winter	Early to mid-autumn	10°C (50°F)
Swede	Late spring to early summer	8°C (46°F)
Sweetcorn	Late spring to early summer	10°C (50°F)
Turnip	Mid-spring to late summer	8°C (46°F)

broadcast sowing, gives a more random effect. It is particularly useful for sowing annual plants once the danger of frost has passed so that they germinate and grow to flowering without having to be transplanted. After sowing the whole area can be raked gently to mix the seed into the soil.

SOWING LARGE SEEDS

If seeds are large enough to handle individually they can be placed in position accurately along a drill, and then covered over. If you are sowing just a few large seeds, such as acorns or beans, you can even use a dibber to make a suitable planting hole.

SOWING SMALL SEEDS

Small seeds are easier to keep track of if you mix them with horticultural sand and sow the whole mixture. Cover them with finely sieved soil and water gently. If you water too strongly, the fine soil may form a hard 'cap' or crust, which the young seedlings will have difficulty in penetrating.

Light seeds can blow away in the wind when they are sown if not held down. If you are sowing a type of seed that does not need covering, this can be difficult, but one way round it is to use a sprinkling of fine grit over the entire surface. This is heavy enough to hold the seeds in place and will allow sufficient light to penetrate between the stones for the seeds to germinate.

SOWING ANNUALS

Summer-flowering annuals can be grown in their flowering position once the danger of a late frost has passed. Prepare the soil as a seedbed for sowing and mark out the drifts for each type of seed with lines of horticultural sand so they are

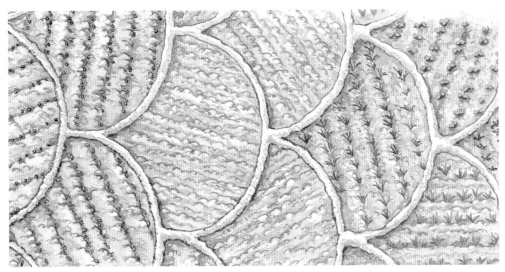

Mark out drifts of different types of plant with horticultural sand or a sharp cane and align the rows in different directions in each section so that the rows will not be apparent when the plants are fully grown.

clearly visible and the seed can be sown within the sand lines. After sowing the seeds, rake each marked area of the seedbed in a slightly different direction to leave the seeds in rows in the bottom of the markings made by the rake's teeth. By raking in slightly different directions, the rows of seedlings are arranged so that it is not possible to see through the bed from one side to the other once the plants have grown up to flower. Also, because the

Preparing to sow

Covering the soil of the seedbed with black plastic a few weeks before you are ready to sow will allow it to absorb the warmth of the sun. This can raise the soil temperature by several degrees before you sow, so that the seedlings get a really good start.

seedlings emerge in rows after germination, it is easier to weed.

THINNING

Just as seedlings in pots and seed trays must be pricked out and transplanted as they develop their first true leaves, seedlings growing in the open ground should be thinned out. No matter how thinly seeds were sown, more seedlings than are needed will usually germinate, and they must be thinned out to ensure that the plants that are left have the space and nutrients to develop and grow strongly. Pests and diseases are more likely to be a problem among plants that are growing too closely together.

Most seed packets indicate the right distance between plants. Remove the unwanted seedlings as soon as possible, leaving the strongest plantlets. Firm the rows gently around the remaining plants and water.

CHILLING

You may find that, no matter what you do to some seeds, they will not germinate until they have experienced a winter outside. One way to overcome this is by the process known as 'chilling', which involves putting the seeds into your refrigerator for a few weeks to simulate a winter outdoors.

Put the seeds in a polythene (plastic) bag, together with some moist sand to stop them drying out and seal the top. Place it in the salad compartment at the bottom of the fridge for between six and eight weeks and then sow the seeds outside in the ground or in pots. As long as the seeds were chilled sufficiently they will germinate.

If you want to chill the seeds of plants such as roses naturally, without sowing them straight away, you can stratify them. Put a layer of damp sand in a screw-top jar, add a layer of seed, then more sand, and so on until the jar is full. Screw the lid back on and half-bury the jar in the garden over winter. The seeds will get their chilling and the sand will stop them drying out.

protecting seeds and seedlings

For the first few weeks of their lives, seedlings are vulnerable to attack and damage. Any extremes of weather may check or even kill the new growth; and, because they do not have the reserves to fight off attacks by pests or diseases, if they are exposed to insect pests or fungal or bacterial disease they are more likely to die than recover. It is up to you to protect them from as much as possible, and this is not as difficult as it might seem. A few simple measures should ensure that the seedlings get the best possible start in life and go on to form strong, healthy, productive plants.

FROST

The weather can be extremely unpredictable and changeable in spring, with warm days and cold nights. The worst conditions as far as seedlings are concerned are clear, cold nights, when the temperature plummets and a frost forms. The water within the plant cells can freeze, expand and rupture the cell walls, causing irreversible damage to the seedling.

A coldframe or unheated greenhouse should offer enough protection to keep the seedlings frost-free during the danger period, but if you do not have the facilities to protect young plants under cover, sheets of newspaper or a piece of horticultural fleece laid over the top of the plants will be sufficient to stop several degrees of frost. Place them over the plants in the

Left: Covering plants with horticultural fleece will protect them while still allowing light to penetrate.

Right: Plastic drink bottles make ideal mini-cloches. The caps can be removed to provide ventilation.

early evening before the forecast frost and lift them off again the following day as the ambient temperatures lift.

WIND
Strong winds can damage plants by scorching the cells of the leaves. Young leaves are particularly vulnerable, even on established plants, but a larger plant will have the reserves of energy to produce more leaves, whereas a seedling will not. Even a strong draught from a window is sufficient to damage tender young growth.

Make sure you position the trays or pots of young plants in a sheltered spot, especially just after germination and while they are being hardened off. Once they are ready to go out into the garden, they will be strong enough to withstand a reasonable wind, but if you live on an exposed site, or if the plants are intended for a windowbox or balcony, you can erect a small screen around them for the first few weeks

using a strip of fleece and some short pieces of bamboo cane to hold it upright.

BRIGHT SUNSHINE
The intense heat of bright, direct sunlight can heat the water within the surface cells of a leaf until it is hot enough to damage them. Once a cell suffers this damage, it dies and no longer contributes to the production of the sugars and starches that the plant needs to survive. Because the damage is seldom restricted to just one cell, the growth of the entire plant can be severely restricted or, in extreme cases, stopped altogether – at which point it will die.

If the seedlings are in a greenhouse or coldframe, the structure can be covered with netting during the day or given a coating of shading paint (use 1 part white emulsion paint to 10 parts water). If the plants are indoors on a windowsill, they will need to be moved away from the direct light during the hottest part of the day or covered with a

single sheet of newspaper. If the trays are standing outside, horticultural fleece laid over the top is less likely to blow away than a sheet of newspaper.

PESTS AND DISEASES
The first line of defence for protecting young plants from pests and diseases is to keep the plants growing as well as possible (see pages 114–17). Once a plant has started to succumb to a particular pest or disease, it becomes vulnerable and is often beset by other problems, which sap the plant's strength even further.

One of the difficult lessons of propagating is to learn to be ruthless. Weak or struggling plants will attract pests and diseases both to themselves and to neighbouring plants, which may have been perfectly healthy to begin with. Check seedlings regularly and remove and discard any that are not growing strongly. Use spray chemicals only as a last resort when all else has failed.

5 DIVISION

Some plants form clumps as they grow rather than having single stems. This habit of growth may be because the plant has spreading under- or overground stems, because it seeds easily, shedding the seeds close to the parent plant, or because it reproduces by having a crown of offset shoots.

Left: Dividing herbaceous perennials seems a very severe treatment for plants, but most will grow better if they are divided every few years or so.

Right: *Erigeron* can be propagated by division during the dormant season and replanted immediately.

Plants that form clumps often become congested and tangled, with the stems competing for the light, moisture and nutrients in a small area. This competition inhibits healthy growth, causing the plants to be smaller and less vigorous and to produce smaller flowers and fruits. For their own sake, such plants need to be thinned out, even if you do not wish to propagate them, in order to maintain their health and their vigour.

Many herbaceous perennials form a clump that may last for several years, but within five years of planting little of the original plant will remain because individual shoots will die after they have flowered and will be replaced by the new shoots that will bear the next year's flowers. This does not apply to the root system, however, because although the top of the plant follows an annual pattern of growth the part below ground is perennial and will last for several years before beginning gradually to die off.

In the wild these plants would spread by growing only around the perimeter of the clump, while the centre would eventually die out. Unfortunately, as this material dies, it begins to rot and starts to attract pests and diseases, which may infect the new, young, healthy growth. Lifting the plants and splitting the large mass into smaller pieces regularly every three to five years, even if they are not being propagated, will rejuvenate them.

You can do this in rotation, rather than having to do all of your perennials together, and those you keep can be replanted, either in the same place if you wish or around the garden. Get rid of all dead or diseased material, keeping only healthy-looking pieces. Any plants for which there is no room can be disposed of or given away. The remaining plants will have the room to grow well again, although you will need to repeat the procedure and split them up again in a few years as the process repeats itself.

There are some types of division that are not quite as straightforward as splitting plants apart and replanting smaller segments into new positions in the garden.

Some plants are so keen to strike out on their own that they will branch out and root from the parent plant, even before they are taken as cuttings. These species are ideal for providing a quick start in the spring because you do not have to wait for the new plants to develop a root system before planting them out. One of the worse problems with dividing plants is that any pieces which break off and stay in the soil will start to grow, usually in places where you do not want them.

Left: *Kerria japonica* 'Pleniflora' has a spreading suckering habit. This makes it ideal for dividing into many new plants.

Right: These *Hemerocallis* 'Burning Daylight' have thick, fleshy roots and can be divided very easily.

Below right: *Eryngium* (sea holly) has a woody crown and a saw will be needed to cut the crown into sections.

how to divide

Lifting and dividing a plant sounds like a simple operation, but when you have dug up a large tangled mass of matted roots and plant, covered in soil, it can be very difficult to know where to start. Make sure you have got the whole plant out of the ground by digging around the edge in a circle and severing the roots underneath so that you

can lift out the whole rootball. Move it to a clear area where you can work easily and lay the plant on a flat surface. So that you can see the roots properly and assess where you will be cutting, wash as much soil off the roots as you possibly can, turning over the plant as necessary so that you can see the entire root system.

ROOT SYSTEMS

When you have cleared the roots of as much soil as possible, look at the rootball and try to see what sort of growth habit it has:
- Thick, fleshy roots (rhizomes) that spread underground – some irises, for example.
- Multiple rosettes of leaves growing together – primroses, for example.

- A mass of leaves arising from a central clump with thin roots below – ornamental grasses, for example
- Multiple woody stems arising from a mass of roots – *Kerria japonica* and *Symphoricarpos* spp., for example

In addition to herbaceous perennials and many shrubs, some plants, such as daffodils, produce lots of small bulbs clustered around a larger one, and others, such as crocuses, produce lots of small, individual corms. These plants are also propagated by division (see pages 94–103).

Each type of root system should be divided in a slightly different way so that the new plants have all that they need to grow when they are replanted. Each type, however, must have some roots and at least one growth bud or quantity of leaves or it will not be able to grow into a new, healthy plant.

DIVIDING LARGE CLUMPS

Plants such as *Carex morrowii* (sedge) and *Hemerocallis* cultivars (daylily) will form huge clumps, which can be difficult to split. The best approach is to dig a trench around the clump and to use a garden fork to prise parts of the plant outwards into the trench. These manageable sections can be divided several times into smaller sections suitable for replanting. This process can be repeated around the clump until all of it has been dislodged and split up.

DIVIDING SMALL CLUMPS

A small clump is usually defined as one that is light enough to be picked up without too much effort by one person, and they are much easier to deal with. The clump can be lifted onto a bench or table and worked on at a convenient height, using two or more forks to divide the plant into smaller sections.

USING FORKS AND TROWELS

One of the easiest ways to divide a mass of fibrous roots is to insert two forks into it, with the tines back to back in the centre, and prise them apart. Depending on the size of the clump, small hand forks may be adequate or you may need to use full-size garden forks. Small clumps can also often be parted with two trowels.

USING A SAW

A few hardy herbaceous plants, such as *Eryngium* spp. (sea holly), have a woody base or crown, which can be difficult to divide, especially if it is one thick stem rather than several smaller ones compressed together. Use an old saw, because the soil particles and stones around the rootstock will soon damage a new one. You could use an axe rather than a saw, but this can be dangerous and lead to high wastage of plants unless you have a good aim.

Left: *Sedum Spathulifolium* will form roots at any point where the stem touches the soil.

Right: Hostas should be lifted and divided in spring.

plants with fibrous roots

Clump-forming and fibrous-rooted herbaceous perennials with a fairly loose crown are propagated by division, which is usually done in spring, just before plants start into growth, so that they have an entire growing season to become established. Plants that flower in spring, however, can be divided in autumn or immediately after flowering, but the newly planted divisions must be well watered throughout dry spells in summer. Propagating plants with thick, fleshy roots presents a number of problems, particularly if the roots are brittle. Roots that snap easily have to be handled with great care, otherwise by the time you have divided the plants into smaller sections there will not be a root in sight, as they will all have broken off. These plants need to be gently teased apart with a minimum of pressure to reduce the chance of bruising or breaking the roots. Removing most of the soil first lets you see how the roots interlock, making it easier to ease them apart.

Plants with a root system that consists of thin, fibrous roots are

HERBACEOUS PERENNIALS

The following plants can be lifted and divided at the time indicated.

Plant	Time
Achillea spp. (yarrow)	Spring or autumn
Agapanthus spp. (African blue lily)	Spring or late summer
Alchemilla mollis (lady's mantle)	Spring
Anemone - hybrida (Japanese anemone)	Spring or autumn
Artemisia spp.	Spring or autumn
Aster spp.	Spring
Astilbe spp.	Spring
Cortaderia selloana (pampas grass)	Late spring
Delphinium cultivars	Spring
Dicentra spp.	Spring
Echinops ritro (globe thistle)	Autumn or spring
Erigeron spp. (fleabane)	Spring
Eryngium spp. (sea holly)	Early spring
Geranium spp.	Spring or autumn
Helleborus cultivars (hellebore)	Spring or late summer
Hemerocallis cultivars (daylily)	Spring or autumn
Hosta cultivars	Spring
Kniphofia cultivars (red-hot poker)	Late spring
Ligularia spp.	Spring
Monarda didyma (bergamot)	Early spring or autumn
Nepeta spp.	Spring or autumn
Ophiopogon spp. (lilyturf)	Spring
Paeonia cultivars (herbaceous peony)	Early spring or autumn
Phlox spp.	Spring or autumn
Rudbeckia spp. (coneflower)	Spring or late autumn
Sedum spectabile (ice plant)	Spring
Stachys byzantina (lamb's ears)	Spring
Tradescantia virginiana	Spring or autumn

comparatively easy to divide, and although the roots may look quite frail and thin, they are usually surprisingly resilient.

BASAL CUTTINGS

Despite the name, basal cuttings are actually a form of division. They are taken early in the season from short, soft shoots that arise from soil level. The shoots are removed from the plant with great care at (or just below) soil level with a sharp knife. With some plants, such as chrysanthemums, the basal cuttings will already have formed roots while they were still attached to the parent plant. These pre-rooted cuttings are traditionally referred to as

Basal cuttings are taken from shoots just as the leaves unfurl in spring. Use a sharp knife to remove a shoot at or just below soil level.

'Irishman's cuttings'. Plants that can be propagated in this way include *Aster* x *frikartii* spp. and cultivars, delphiniums, *Nepeta* spp. (catmint) and *Stachys* spp.

Dividing plants with fibrous roots

When you are dividing plants that have thin, fibrous roots, such as asters, it is important to work as quickly as possible so that the plants do not deteriorate. The fine roots dry out quickly and areas of root may die – which means that the chances of the newly planted divisions surviving are reduced.

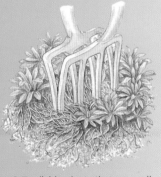

1 *Start by lifting all or the part of the plant that is to be divided with a spade or fork. (A fork usually causes less damage to the roots.) Large clumps may need to be lifted in several sections. If the soil is heavy, wash the roots with a hosepipe to get rid of as much soil as possible.*

2 *To divide plants that are well established and have a dense, thickly matted rootball, a pair of garden forks can be placed back to back and driven through the rootball. Pulling the fork handles together will split the rootball, allowing the clump to separate into two smaller sections.*

3 *Keep only the newest, most vigorous shoots and always discard any old or decaying sections of plant. These sections can harbour pests and diseases and transfer them to the healthy shoots. Replant the sections of plant that are to be kept, with the roots just below the soil surface.*

Left: Many penstemons spread over the ground, rooting as they go. These sections can be cut from the parent plant and replanted.

Right: Bamboos can be propagated to stop them spreading. Digging up the spreading suckers and potting them keeps the plant under control.

woody plants and clump-forming shrubs

Many shrubs that are multi-stemmed or have a suckering habit can be increased by means of division.

Sometimes only a small section of a shrub is split from the parent plant. It can be chopped from the rest of the plant with a sharp spade or, in the case of *Fargesia nitida* (fountain bamboo), an axe. Small segments of the root of some of these shrubs that have been severed from the main plant may also develop into new plants as these sections become, in effect, root cuttings (see pages 86–7).

MULTI-STEMMED AND SUCKERING SHRUBS

Plant	Time
Cornus stolonifera (red osier dogwood)	Early winter
Deutzia scabra	Autumn
Fargesia nitida (fountain bamboo)	Spring
Hypericum calycinum (rose of Sharon)	Spring or autumn
Kerria japonica	Autumn
Leycesteria formosa (Himalayan honeysuckle)	Autumn
Lonicera fragrantissima (honeysuckle)	Autumn
Penstemon spp.	Spring
Potentilla fruticosa	Spring or autumn
Rubus thibetanus	Autumn to spring
Spiraea japonica	Late autumn or early spring
Symphoricarpus orbiculatus (coralberry)	Autumn

When sections of these plants are divided from the main plant it is advisable to remove at least the top half of the growth on the new plant so that it does not rock in the wind when it is has been transplanted. Any movement of the topgrowth will prevent the roots from spreading into the surrounding soil to become established.

SUCKERS

New shoots that grow directly from the roots of a plant at some distance from the previous main stem are known as suckers. Many of the plants that can be propagated by division spread naturally because of their ability to produce these new shoots from roots that spread just below soil level. These roots can spread for long distances under the soil, so that these suckering shoots can emerge several metres away from the parent plant.

A sucker can be severed from the main plant using a spade, before being dug up and transplanted into a new position. However, any sections of severed root that are left in the ground after the plant has been taken out also have the capacity to produce new plants if they have any dormant buds attached to them. This may lead to a lot more small suckers being produced in the area where the main one has been dug up – and they may not always be welcome so think carefully before you do this.

Root bruising

Plant roots that have been cleanly cut usually heal quickly with no extra help, but bruised root tissue is slow to heal and susceptible to rotting, so always prune bruised roots back to healthy tissue before replanting. As a precaution, you can dust the roots lightly with sulphur to act as a natural fungicide and help eliminate any root rot before it becomes established.

INVASION

As well as being a simple and easy method of propagation, division can also be used as a method of keeping invasive plants, such as bamboo, under control. This may involve lifting the entire plant and dividing it at regular intervals or working around the base of the plant and separating small sections before they start to spread. The real advantage of this technique is that not only are you controlling the growth of the plant but you are also gaining extra plants.

rhizomatous plants

Some herbaceous perennials develop substantial rhizomatous root systems, and these plants are also propagated by division. Rhizomes are actually modified stems, which grow more or less horizontally at or near to the soil's surface and from which the shoots arise and roots descend. In some species the rhizome stores food. Most irises, including bearded irises, grow from rhizomes, the exceptions being the bulbous irises in the Reticulata, Xiphium and Juno groups. Rhizomatous perennials are usually propagated after flowering, when the rhizome naturally produces new roots. After lifting, the rhizome can be washed to remove the soil. It can then be cut into sections.

Replanting

After a plant has been divided it is important to take great care over replanting the small divisions. If they are planted shallowly it is likely that they will establish only slowly – or even die – because lack of water causes them to dry out. However, far more newly planted herbaceous perennials die because they are planted too deeply and fail to grow as they have begun to rot because there is too much water and too little oxygen around the roots.

CROWN RHIZOMES
Some plants, such as some peonies and asparagus, develop rootstocks that can be divided into clumps. When the rootstock

is divided each piece must have at least one well-developed bud. Crown rhizomes should be divided in late winter or early spring with a garden spade or sturdy knife. If the rootstock is divided into two or three pieces they can be replanted immediately. Several smaller divisions can be potted up individually and grown on for a year before being planted out. All cut surfaces should be dusted with a fungicide to prevent rot.

RHIZOMATOUS PLANTS

Plant	Time
Anemone altaica, A. eranthoides	Spring
Begonia rex and rex cultorum hybrids	Summer
Bergenia cordifolia (elephant's ears)	Spring or autumn
Canna cultivars (Indian shot plant)	Early summer
Convallaria majalis (lily-of-the-valley)	Early spring
Iris spp. (some)	Summer
Mentha spp. (mint)	Growing season
Ophiopogon planiscapus (lilyturf)	Early spring
Polygonatum biflorum (Solomon's seal)	Early spring
Smilacina racemosa (false Solomon's seal)	Spring
Zantedeschia aethiopica cultivars (arum lily}	Spring

Dividing rhizomatous perennials

Plants with rhizomatous roots, such as the bearded iris Iris germanica*, are best divided in the summer immediately after they have finished flowering.*

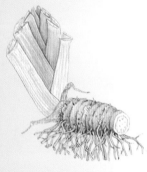
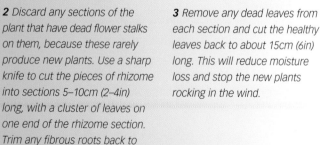

1 Use a fork to lift the rhizomes and wash off as much soil as possible, before pulling the clump apart to get as many sections as possible. Each section needs a 'fan' of leaves and a piece of the rhizome.

2 Discard any sections of the plant that have dead flower stalks on them, because these rarely produce new plants. Use a sharp knife to cut the pieces of rhizome into sections 5–10cm (2–4in) long, with a cluster of leaves on one end of the rhizome section. Trim any fibrous roots back to about half their original length.

3 Remove any dead leaves from each section and cut the healthy leaves back to about 15cm (6in) long. This will reduce moisture loss and stop the new plants rocking in the wind.

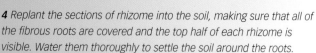

4 Replant the sections of rhizome into the soil, making sure that all of the fibrous roots are covered and the top half of each rhizome is visible. Water them thoroughly to settle the soil around the roots.

Left: Strawberry runners are horizontal stems that spread over the ground to produce new plants.

offsets, runners and plantlets

Plants have a number of methods of spreading and propagating themselves, and these have developed as a means of taking over and colonizing an area to reduce the competition and to make conditions more favourable. In nature, the negative side of this is that, because all the plants are extracting the same nutrients from the soil, the stocks eventually become depleted and the plants suffer. In the garden, where the plants can be fed regularly, this is not a problem, so the difficulty becomes trying to keep these plants within the boundaries you have set. Regular division will help to control their spread and provide extra plants for elsewhere.

OFFSETS

An offset is a new plant that grows from the base of the parent, usually where the parent has a basal rosette of leaves, such as cordylines and yuccas, rather than a single woody stem. As soon as the new plant is large enough to support itself, it can be detached and potted up to grow separately. Leaving them in place means that the offsets may eventually begin to crowd the parent plant.

Some rock garden plants produce offsets as a means of survival. The main parent rosette of plants such as saxifrage and sempervivum (houseleek) will die after it has produced a stem and flowers. The plant will try to reproduce sexually by forming and shedding seeds, and, as a back-up, it will reproduce vegetatively by sending out modified horizontal stems. These stems can spread above or below ground, depending on the species, but the tip of the

Rosette-forming plants

Some rosette-forming plants, such as saxifrages and sempervivums, may accumulate dead leaves at the bottom of the rosette. Whenever these plants are propagated, these leaves must be removed. If they are left they often dry and curl up, raising the rosette up from the soil or compost, which causes the whole rosette to dry and die.

Strawberry runners can be pegged into compost-filled pots or direct to the ground. When the runners have developed good roots they can be severed from the parent plant.

modified shoot will stop growing away from the parent plant after a period and form a new, independent plant with another rosette. These rosettes can be severed from the parent plant and planted out to grow independently. The horizontal stem should be kept and cut to a length of 2–3cm (1in) to act as an anchor for the rosette, and it will often produce roots to help the plant grow.

RUNNERS

Some plants develop horizontal stems that produce a number of young plants along their length. One of the most widely grown plant that produces runners is *Fragaria* spp. (strawberry), which can produce runners that will grow for 2m (6ft) or more in a single season.

Runners grow along the soil's surface and are usually weighed down by the plantlets as they form along its length. Where the plantlet comes into contact with the soil it will form roots, which

then grow down and anchor the plantlet. Eventually the runner stem withers and dies. The process can be speeded up by pegging down the plantlets with heavy-gauge galvanized wire so that the plantlets root more quickly.

If a strawberry runner is cut off just beyond the first plantlet after it has started to root, fewer (but larger) strawberry plants will be produced. These will usually yield a heavy crop of fruit in their first year of growth.

PLANTLETS

Some plants are capable of producing embryo plants (adventitious plantlets) on the edges or on the flat blade of their leaves. Succulent plants, such as some forms of *Sedum* and *Kalanchoe daigremontiana*, will shed these plantlets from the leaves, and many will root into the sand on a greenhouse bench or into the compost around the base of the parent plant. Other plants, like *Asplenium bulbiferum* (hen-

Some ferns, including *Asplenium bulbiferum,* can be propagated by laying fronds on compost in a seed tray. Plantlets develop on the frond, and when they are about 5cm (2in) high they can be lifted and potted up.

and-chicken fern) and *Tolmiea menziesii* (pick-a-back plant), can be propagated easily by cutting off leaves from the parent plant and pegging them down onto a tray of compost. Plantlets will grow on the upper side of the leaf and their roots will grow into the compost from the undersides of the leaf.

6 CUTTINGS

A cutting is a piece of plant material that has been removed from the parent and that has no means of supporting itself and is, therefore, vulnerable during the early days until it can form its own root system. It is the task of the propagator to help it do this as quickly as possible and with the minimum stress to the plant. The new plant should be identical to the parent in every way, although some small variations have been found to occur, and this is often how new varieties and cultivars are developed.

Left: *Forsythia* is very easy to grow from any type of stem cutting which may explain why it is so popular.

Right: *Buddleja* 'Dartmoor' will propagate very easily from hardwood cuttings about the size of a pencil.

The price of a plant in the garden centre can often be a reflection of its ease of propagation or speed of growth. The more expensive the plant, the more difficult it is to grow or the more slowly it grows. When you are starting to propagate, the expensive plants are usually those you should avoid, although they may eventually be the ones that give you the greatest satisfaction if you ever manage to root them.

Begin by taking a good look at the plant you are going to use, and identify the flowering shoots and the growth shoots. The flowering shoots will have either flowers or buds showing at the tip or along the stem itself, even if the flowers are dying off. The buds on the growth shoot will be smaller and thinner. It is important to distinguish between the two types of shoot because the balance of plant hormones within each one is different. In a flowering shoot it is tipped in favour of producing flowers and, ultimately, seeds. In a growth shoot it is aimed at growth, which is exactly what it needs if it is going to produce roots.

Once you have found a growth shoot, look closely at the shoot itself. As you move down the stem away from the small leaves that surround the bud at the tip, the leaves will be arranged either singly or in pairs. Where they are joined to the stem there is a slight swelling, known as the leaf joint or node. The chemicals (hormones) within the plant, which are needed when roots are being produced, are concentrated in certain parts of the shoot. The greatest concentration is found in the growing tip and lesser concentrations occur at the nodes. When you prepare a cutting, therefore, by trimming it just below one of these leaf nodes you are making sure that the chemicals are as close as possible to the area where they will be needed for the roots to grow.

taking cuttings

There are several types of cutting, which make use of various parts of the parent plant, including leaves, shoots and roots. Those taken from shoots include tip cuttings, which have the growing tip and two or three leaf nodes below it; stem cuttings, which do not have a growing tip but are taken by cutting above one leaf node and below another; and leaf bud cuttings, which consist of a single leaf and its node, cut above and below. Root cuttings, which are the only ones taken from below ground, involve cutting pencil-thick roots into small pieces.

TIMING

Outdoors, different cuttings are taken at different times of year, according to the growth of the plant and its stage of maturity.

Root cuttings, for example, are taken in winter when the plant is at its most inactive and will not mind the disturbance (as it would in summer when it was actively growing). Soft tip cuttings are usually taken in early summer, when the growth tip has fully developed but has not yet started to turn woody and hard. Semi-ripe cuttings are taken in late summer and are either tip or stem cuttings with some brown (ripe) tissue at the base, which has started to harden and turn woody. Hardwood cuttings are taken in late autumn or early winter from the bare stems of deciduous shrubs, and they root slowly outdoors over winter and spring. Houseplants can be propagated in a number of ways, just like outdoor ones, but they are less dependent on the

season. Some can be divided, others send out runners, but many can be grown from one form of cutting or another. Plants such as some types of begonia,

Easy cuttings

The following plants are especially easy to propagate from cuttings:

- *Buddleja* spp. (buddleia)
- *Chrysanthemum* cultivars
- *Delphinium* cultivars
- *Forsythia*
- *Fuchsia* cultivars
- *Pelargonium* cultivars
- *Ribes* spp.
- *Rosa rugosa*
- *Tradescantia* spp.
- *Weigela* spp.

Left: Growing young plants in peat pots makes transplanting very easy. The pot can be planted out with the plant.

Right: Fuchsias root very easily and quickly. They can even form roots in a jar of tap water.

for example, can be propagated from leaf cuttings, which involves cutting a leaf from the parent plant and using it to create one or more new plants.

SELECTING MATERIAL

Once you have identified the plants that are ready for use as cuttings material you will have to identify the individual shoots to cut. As long as you have a reasonable amount of material to choose from you can be quite selective about what you take. Use the healthiest material available – look for sturdy stems and firm leaves and make sure there are no signs of damage from pests or diseases. Do not use anything with any fungus or mould on it, as this will spread through the batch in no time. Similarly, leaves that show signs of having been nibbled should be avoided in case the pest has left eggs behind that will hatch out and eat their way through the

batch of fresh cuttings.

Do not take flowering shoots, because the balance of hormones in the shoots is tipped in favour of flowering and fruiting rather than growth (see pages 70–71). Check there are no flower buds developing.

COLLECTING MATERIAL

As you collect material for cuttings, bear in mind that you are removing parts of the growing plant. Try not to remove so much that the plant is prevented from flowering or is disfigured, and look for suitable shoots from around the back of the plant. Aim to collect cuttings in the morning, before the day gets too warm, because if the plant cells are still plump with overnight moisture they will be in a better condition to withstand the period when they will be without roots. Ideally, cuttings should be inserted into compost as soon as they have been

collected, but if you are not going to prepare the cuttings straight away put the material into a plastic bag containing a tiny amount of water. Fold over the top loosely and stand it in a cool, shady spot until you are ready.

Types of cutting

Cuttings are identified by the part of the plant that is used as the cuttings material. The four types are:

- *Tip cuttings, which include soft tip, semi-ripe and heel cuttings.*
- *Stem cuttings, which include hardwood, nodal and internodal cuttings.*
- *Leaf cuttings, which include leaf bud, leaf petiole and whole leaf cuttings.*
- *Root cuttings.*

Left: Large cuttings of *Pyracantha* will produce flowers and fruit within a year of being rooted.

Right: *Impatiens* will root very quickly if they are given plenty of water.

rooting cuttings

Until a cutting has developed its own roots, it has no support system to keep it alive. The moisture and nutrients that were previously supplied by the parent plant now have to be supplied by the propagator until the cutting has produced some roots of its own, and is able to maintain itself.

Large cuttings

Some plants root easily, and you can take much larger cuttings than normal if you have the facilities to root them. Pyracantha spp. (firethorn), for instance, will grow from cuttings up to 45cm (18in) long, and each plant will yield fewer, but much larger, cuttings.

PREPARING CUTTINGS

Try not to take the cuttings until you are ready to insert them into the compost, or the cut surface will dry out. Before you even cut off the shoots fill the pot or seed tray with compost and press it lightly to firm it. Prepare the cuttings as a batch and insert them straight away.

The length of the cutting will depend on its ability to survive without roots. A woody cutting can survive longer than a soft, sappy one, which will wilt quickly. About 8cm (3in) is a good average, but cuttings of up to 1m (3ft) will root for some forms of easy-to-root plants, such as forsythia.

Make the cut at the appropriate point on the stem, using a clean, sharp knife or a pair of secateurs. Strip away the lower leaves, which would be in contact with the compost, because these would quickly rot and could infect the cutting and those around it.

INSERTING THE CUTTINGS

Whether you use a pot or seed tray for the cuttings, fill it with a seed and cuttings compost, tap it to get rid of any air pockets and level the top to remove the surplus compost. Firm the compost lightly; do not compress it, however, or the roots will have trouble penetrating it. Prepare a batch of cuttings of a single type, and insert them into the compost together. Work along one side and down another to work out the spacing in a tray or position them around the perimeter of a pot. Push the cuttings down into the compost, leaving the lowest leaves above the surface so that they have no contact with the compost, and gently firm them in with your fingers.

AFTERCARE

Once they have been inserted into the compost, the cuttings should be placed in a warm, moist environment. Extra moisture in the air around them will prevent them from losing the water they contain within their cells, and this will stop them from wilting.

If you have a propagator, place the filled tray or pot inside so that you can control the humidity and provide some bottom heat. Alternatively, use a plastic bag. A tray can be placed inside a large bag and the end sealed or a small bag can be used for a single pot, either with the pot sealed inside or the bag held over the top of the pot with an elastic band. It is important that the plastic does not touch the leaves on the cuttings, so support the plastic bag on wire hoops or sticks.

Check the cuttings regularly for any fungal infection, which will show itself as a furry mould. Remove and destroy any cuttings that are infected as soon as you notice them.

Wounding

A shallow cut, or wound, can be made along the bottom 1cm (½in) of the cutting to help encourage roots to form in that area.

Adventitious roots

Some plants will root from cuttings without using compost at all. Many herbs, including mint, and also fuchsias and some types of Impatiens (busy lizzie), will readily form roots from tip cuttings suspended in a jar of tap water. When the roots are 3–5cm (1–2in) long, carefully pot up each cutting into an individual pot.

Left: Non-flowering shoots of *Hydrangea macrophylla* will root readily as softwood cuttings taken in the summer.

tip cuttings

Once you have prepared each cutting (see pages 72–3) it should be inserted into the compost as quickly as possible and moved to the propagation environment where the humidity can be controlled and the amount of moisture being lost from the leaves reduced to an absolute minimum. For the rooting process to be successful, it is important for the cutting to be in as close contact with the compost as possible on all sides. For this reason, it is usually better to push the cutting into lightly firmed compost rather than make a hole with a dibber or pencil and insert it. Any air pockets that are left around the cutting will allow that part of the cutting to dry out rather than form roots.

Long cuttings

If tip cuttings are too long, always shorten them from the base upwards rather than removing the growing point, as this is where they produce the plant hormones that encourage rooting.

SOFT TIP CUTTINGS

A soft tip cutting includes the sappy tip of a growing shoot plus

SOFTWOOD CUTTING

Plant	Time	Plant type
Abelia spp.	Summer	Deciduous or evergreen shrub
Aphelandra spp.	Summer to autumn	Houseplant (evergreen shrub)
Begonia	Spring to summer	Houseplant (all types)
Callicarpa spp. (beauty berry)	Spring	Deciduous or evergreen shrub
Campanula isophylla (Italian bellflower)	Early spring	Houseplant (perennial)
Chaenomeles spp. (flowering quince)	Summer	Deciduous shrub
Clematis spp.	Spring	Deciduous climber
Codiaeum spp. (croton)	Spring to summer	Houseplant (evergreen shrub)
Crassula spp.	Spring to summer	Houseplant (evergreen shrub)
Dieffenbachia spp. (dumb cane)	Spring to summer	Houseplant (evergreen perennial)
Ficus spp.	Spring to summer	Houseplant (evergreen shrub)
Forsythia	Late spring to early summer	Deciduous shrub
Hoya carnosa (wax plant)	Late summer	Houseplant (evergreen climber)
Hydrangea spp.	Summer	Deciduous shrub
Hypericum spp.	Summer	Deciduous or evergreen shrub
Impatiens spp.	Spring to early summer	Houseplant (perennial)
Justicia brandegeeana (shrimp plant)	Late spring	Houseplant (evergreen shrub)
Kalanchoe spp.	Spring or summer	Houseplant (perennial)
Monstera deliciosa (Swiss cheese plant)	Summer	Houseplant (evergreen climber)
Peperomia spp.	Spring to summer	Houseplant (evergreen perennial)
Saintpaulia culticvars (African violet)	Spring to summer	Houseplant (evergreen perennial)
Stephanotis floribunda	Early summer	Houseplant (evergreen climber)
Tradescantia fluminensis	Spring to summer	Houseplant (evergreen perennial)

Tip cuttings

Take tip cuttings early in the day, when the shoots are full of moisture.

1 *Tip cuttings are usually made from the youngest, soft, sappy growths at the end of the current season's growth. Always choose a shoot with a strong, healthy-looking tip and make your cut just below a node (leaf joint) at least 10–13cm (4–5in) below the tip with a sharp knife or secateurs.*

2 *Trim each cutting to 8–10cm (3–4in), cutting just below a leaf joint with a sharp knife. Remove all the leaves from the bottom third of the cutting (which will be buried in the compost) by pulling them gently downwards. This will leave small wounds from which roots will emerge.*

a short section of stem. In mid- to late spring select a healthy-looking new shoot and remove the growing tip, making the bottom cut just below a leaf joint (node) to leave a cutting 8–10cm (3–4in) long; this will, of course, vary slightly according to where the leaves are positioned. Remove the leaves from the lower third by pulling carefully or trimming them with a knife, as they will rot in contact with moist compost. If you are using a hormone rooting preparation (see pages 32–3) apply it at this stage. Not every plant needs rooting preparations, especially when using the soft tip cuttings, which tend to root quite easily.

SEMI-RIPE CUTTINGS

As the season progresses, the stems of new shoots begin to turn brown and woody, particularly on shrubs, and some plants root better if the cuttings are taken with just a little bit of this browning (lignification) at the base. Choose healthy-looking growth tips, as before, and leave the tip intact. Ideally, you should end up with a cutting of a similar size to the soft tip – that is, 8–10cm (3–4in) long – with 1cm (½in) of brown wood at the base, cut just below a leaf node. Remove the leaves from the lower third, and dip the bottom cut edge in rooting hormone (see pages 32–3).

Many cuttings can be removed from the parent plant by pulling them from the stem with a small segment of the main stem – a heel – still attached. This open wound often produces a better root system than when the base of the cutting is cut with a knife.

Semi-ripe cuttings

This is an easy method of propagation for a wide range of popular plants, including most conifers and many broad-leaved evergreens. The critical part of the cutting is at the base, where the semi-ripe section of growth is at an intermediate stage between soft- and hardwood. This area is where the bark on the stem has started to change colour from green to light brown.

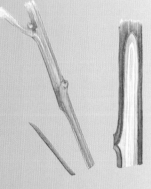

1 Remove shoots 13–15cm (5–6in) long from the branches of the parent plant (the exact length will vary depending on where the bark is changing colour). Trim the base of the cutting to just below a leaf joint or node, before removing the leaves from the lower third of the cutting. If these leaves are pulled off, they will leave small wounds near the base of the cutting, and these become the sites for roots to emerge.

2 Increase the chances of rooting by wounding the base of the stem by cutting away a sliver of bark, about 2.5cm (1in) long. Dip the base of each cutting in hormone rooting powder, if appropriate.

3 Insert each cutting into a pot of cutting compost, arranging them around the edge of the pot so that the leaves do not touch. Water the cuttings thoroughly and place a plastic bag over the pot to reduce evaporation.

SEMI-RIPE CUTTINGS

Plant	Time	Plant type
Actinidia spp.	Late summer	Deciduous climber
Aucuba japonica (spotted laurel)	Summer	Evergreen shrub
Azara spp.	Mid- to late summer	Evergreen shrub
Berberis spp.	Summer	Evergreen or deciduous shrub
Brachyglottis spp.	Summer	Evergreen shrub
Buxus spp. (box)	Summer	Evergreen shrub
Camellia spp. and cultivars	Late summer to late winter	Evergreen shrub
Ceanothus spp. and cultivars	Mid- to late summer	Evergreen shrub
Chamaecyparis spp. (cypress)	Late summer	Evergreen conifer
Choisya spp. (Mexican orange blossom)	Summer	Evergreen shrub
Cistus spp. (rock rose)	Summer	Evergreen shrub
Cupressocyparis spp.	Late summer	Evergreen conifer
Cupressus spp. (cypress)	Late summer	Evergreen conifer
Cytisus spp. (broom)	Late summer	Evergreen or deciduous shrub
Elaeagnus spp.	Summer	Evergreen shrub
Escallonia cultivars	Late summer	Evergreen shrub
Euonymus spp.	Summer	Evergreen shrub
Fatsia japonica	Early summer	Evergreen shrub
Forsythia spp.	Late summer	Deciduous shrub
Garrya elliptica	Summer	Evergreen shrub
Hebe spp.	Late summer to early autumn	Evergreen shrub
Hedera spp. (ivy)	Summer	Evergreen climber
Ilex spp. (holly)	Late summer to early autumn	Evergreen tree or shrub
Itea ilicifolia	Mid- to late summer	Evergreen shrub
Itea virginica	Late summer	Deciduous shrub
Kerria japonica	Early to midsummer	Deciduous shrub
Mahonia	Mid- to late summer	Evergreen shrub
Olearia spp. (daisy bush)	Summer	Evergreen shrub
Osmanthus spp.	Summer	Evergreen shrub
Phlomis spp.	Summer	Evergreen shrub
Pittosporum spp.	Summer	Evergreen shrub
Potentilla fruticosa cultivars	Late summer	Deciduous shrub
Prunus laurocerasus	Midsummer	Evergreen shrub
Pyracantha spp.	Mid- to late summer	Evergreen shrub
Rhododendron spp.	Late summer	Evergreen shrub
Solanum spp.	Summer to early autumn	Evergreen shrub or climber
Spiraea spp.	Summer	Deciduous shrub
Taxus spp. (yew)	Late summer to early autumn	Evergreen conifer
Thuja spp. (arborvitae)	Late summer	Evergreen conifer
Wisteria spp.	Late summer	Deciduous climber

HEEL CUTTINGS

Some plants produce short sideshoots, which are ideal for taking as cuttings. Depending on the stage of growth during the season, they may or may not have developed the brown woody tissue at the base. Pull these shoots gently from the main stem so that they come away with a 'heel' (a tiny section of bark). Remove the leaves from the bottom third, trim the heel to get rid of any long tail it might have and dip it in rooting hormone.

Left: Shrub and species roses, like this *Rosa gallica* 'Versicolor' will grow from stem cuttings and form a spreading, suckering shrub.

Below left: *Weigela florida* 'Foliis Purpureis' will grow from stem cuttings and can be propagated at any time of the year.

stem cuttings

This is the most popular and easiest method of propagation. Stem cuttings are similar to tip cuttings in that each cutting must be inserted into compost as soon as possible so that the cut tissue does not begin to dry out, and so that the cuttings develop roots quickly. Pruning the parent plant hard back will encourage it to produce fast-growing shoots, which are ideal for use as stem cuttings.

HARDWOOD CUTTINGS

A hardwood cutting, taken in late autumn or early winter, is the easiest way to propagate deciduous trees and shrubs. Hardwood cuttings can be taken from evergreen trees and shrubs at the end of the growing season. Take cuttings that are about the thickness of a pencil. If you are propagating shrubs dip the base of each cutting into a rooting powder before burying two-thirds

of their length in the garden soil. Cuttings for trees can be buried with their full length in the soil, so that only the top bud of the cutting is showing (this will encourage a single, straight stem to grow from each cutting).

In heavy clay soils bury the cuttings in a mixture of soil and sharp sand to improve drainage. Inserting the cuttings into the soil through a layer of black plastic is a useful way of reducing the

HARDWOOD CUTTINGS

Plant	Time	Plant type
Buddleja davidii	Autumn	Deciduous shrub
Cornus spp.	Autumn	Deciduous shrub
Deutzia spp.	Autumn	Deciduous shrub
Fallopia baldschuanica	Autumn	Deciduous climber
Forsythia x *intermedia*	Autumn	Deciduous shrub
Juniperus spp.	Autumn	Evergreen conifer
Leycesteria formosa	Late autumn to early winter	Deciduous shrub
Lonicera spp.	Autumn	Deciduous climber
Metasequoia glyptostroboides	Winter	Deciduous conifer
Platanus spp.	Winter	Deciduous tree
Populus spp.	Winter	Deciduous tree
Ribes spp.	Autumn	Deciduous shrub
Rosa spp.	Autumn	Deciduous shrub or climber
Salix spp.	Winter	Deciduous shrub or tree
Sambucus spp.	Winter	Deciduous shrub or tree
Symphoricarpos spp.	Autumn	Deciduous shrub
Taxodium spp.	Autumn	Deciduous tree
Vitis spp.	Late winter	Deciduous climber
Weigela spp.	Autumn to winter	Deciduous shrub

amount of time spent watering or weeding the cuttings. The plastic blocks out the light, which stops weed seeds from germinating, and reduces water loss through evaporation from the soil around the cuttings.

HEDGING PLANTS
Cuttings or seedlings used for hedges can be grown in a small space by using a strip of black plastic 30cm (1ft) wide and 2m (6ft) long. Lay out the plastic and place the cuttings or seedlings at 30cm (1ft) intervals along the edge of the plastic. Sprinkle compost over the middle of the plastic. Fold the plastic over the base of the cuttings or seedlings to form an envelope 15cm (6in) deep; the compost will be at the bottom. Roll up the plastic so that you finish with a circular bundle of plants in a large, black plastic 'pot'.

Hardwood cuttings

The cuttings will do best in a site that is out of direct sun and is sheltered from cold, drying winds. Before inserting the cuttings in the ground, dig over the soil thoroughly and make a V-shaped trench, with one sloping side and one vertical side.

1 *Take the cuttings and remove the soft tips, cutting just above a leaf node. Trim the cutting just below a node to about 15cm (6in) long.*

2 *If wished, dip the base of the cutting in hormone rooting powder, then insert the cuttings in the trench and fill it in. Make sure that the soil never dries out.*

Take a nodal cutting by making a straight basal cut about 3mm (⅛in) below a leaf node. Remove the lowest pair of leaves.

these particular positions are quite practical.

At the junction of each node is a dormant bud that has the potential to develop into a stem. These dormant buds also have the capacity to make cells divide, because they can manufacture the plant hormones that are responsible for cell division. By cutting the stem close to the bud, the plant will respond quickly to heal the wound (and produce roots) by manufacturing these growth hormones.

On the internodal cutting, which is most often used for propagating fuchsias and clematis, the basal cut is made halfway between nodes.

NODAL CUTTINGS

Most cuttings can be described as 'nodal' when the cuts that determine the length of the cutting are made just above a node at the top and just below a node at the bottom. The reasons for siting the cuts on the stems in

INTERNODAL CUTTINGS

Some plants are able to produce healing tissue from almost any part of the stem, which means there is less need for precision when making cuts on the stem because specific points do not have to be chosen. Plants such

as fuchsias, clematis and *Lonicera* (honeysuckle) will often root just as easily when they are propagated from internodal cuttings as from nodal cuttings. It is not uncommon for these plants to produce roots along the entire section of the cutting that

Poor rooting of cuttings

It is tempting, as you insert newly prepared cuttings into the compost, to believe that the next time you see the bottom of the cutting it will be a mass of healthy roots. Sadly, that may not be the case. Many cuttings rot or dry out at the base before the roots get a chance to grow. There are several reasons for this.

- *Too much rooting powder can burn the tissue at the base of the cutting, damaging the tissue so that it cannot form roots. Use the bare minimum and knock off the surplus.*
- *Insufficient rooting hormone will mean that, if the cutting is not able to form roots without an extra boost and this is not applied (by you in the form of rooting powder), it will wither and die.*
- *Taking a flowering shoot instead of a growth shoot as a cutting will cause rooting to fail because the balance of hormones within the cutting is wrong, or because it takes so long before roots form that the reserves within the cutting are exhausted.*
- *Excess watering will cause the cutting to rot because it deprives the cutting of air so that it cannot root.*
- *Too little water will allow the cutting to dry out before it can root.*
- *The ideal temperature for rooting most cuttings is between 13 and 18°C (55–64°F). If it is above 21°C (70°F) for any length of time rooting is slowed down rather than speeded up, and at this temperature most conifers form only callus (scar tissue) rather than roots.*
- *The level of humidity around the cuttings is important. The drier the air, the more quickly the cutting will use up its own reserve of moisture. Until it has roots, this moisture cannot be replaced and the cutting will dry out and, probably, die.*
- *Cuttings with a particularly woody base may need more than rooting hormone because the roots may not be able to get through the thickened bark layer. Wounding the cutting by removing a slice from one side of the stem gives extra surface for the roots to grow and breaks through the tough layer.*

Right: Many clematis will root from internodal cuttings, with just two buds being sufficient to make a new plant.

has been inserted into the cuttings compost.

This capacity to root at almost any point has considerable advantages when you are taking cuttings because only one node is needed on each internodal cutting, rather than the two nodes required on a nodal cutting. This makes it possible to get twice the number of cuttings from the same amount of propagation material.

SMALL CUTTINGS

When you acquire a new plant you may want to take as many cuttings from it as possible. This might be for economic reasons – to get as many plants as possible for the least cost – or because it is a brand-new cultivar, which you want to increase. If you buy a small fuchsia early in the season, for instance, you can take tip cuttings regularly as the plant grows, so that by the summer you can have 10–12 plants. Fuchsias root so easily that they will grow from internodal cuttings, involving just one leaf joint and a short piece of stem. However, if you are careful, you can split this cutting down the centre, leaving a leaf and bud on each piece, and root both. Note, however, that this method works only where the leaf buds are opposite each other on the stem.

Soft fruit

Propagate currants and gooseberries by hardwood cuttings, taking longer cuttings if the parent plant is reluctant to develop roots. The redcurrant (left) and gooseberry (centre) cuttings are similar in length, but leaves are left at the tip of the gooseberry to aid the production of roots. All but four buds should be removed from redcurrant and white currant cuttings to avoid suckers forming, but all the buds on blackcurrants (right) can be left in place.

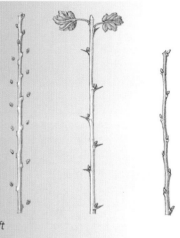

Mallet cuttings

If plants are particularly susceptible to rot, mallet cuttings, which have a large, woody plug at the base of the cutting, are most likely to produce roots. The cuttings, which should be of semi-ripe or hardwood, can be taken from Berberis spp. in autumn and planted in cold frames. Cut the stem just above a sideshoot and then about 2cm (¾in) below the first cut. Remove the leaves from the bottom of the sideshoot and insert the cutting in soil.

Left: African violets (*Saintpaulia*) never really form a stem but the leaf stalk 'petiole' is capable of forming roots and producing new plants.

Some plants can reproduce from small segments consisting of a small section of stem holding a single leaf and dormant bud. Start by making a cut just above the bud and leaf, before making another flat cut 2–3cm (about 1in) below the bud and leaf. Discard the soft sappy tip of the shoot as this will usually rot.

leaf cuttings

Some plants will grow from small sections of stem or even leaves, so you can take a large number of cuttings from a single plant without doing it any great damage. Leaf cuttings can be taken using the entire leaf plus its petiole (stalk), the leaf alone or sections of leaf. A leaf bud cutting includes a short section of stem, a single leaf and its axillary bud (the growth bud found in the angle between the leaf and the stem). Each method will yield different numbers of young plants, and each is suited to a different type of plant.

LEAF BUD CUTTINGS

Plants such as the winter-flowering mahonia, with its holly-like leaves and lily-of-the-valley-scented flowers, will grow from just a single leaf cutting with a short section of stem below. Take the cuttings in summer from the new season's growth and look for a healthy leaf with a good bud. The leaf will feed the cutting while it roots and the bud will form the new stem.

Make one cut 2–3cm (about 1in) below the leaf node and the other just above the node but not so close that it interferes with the bud itself. Large leaves can be rolled up and held with an elastic band or reduced by half, because the cutting does not need all of it.

Dip the bottom cut into rooting hormone and knock off the excess before inserting the cutting into a pot or seed tray of compost so that the bud is level with the surface.

LEAF AND PETIOLE CUTTINGS

A single *Saintpaulia* (African violet) plant can be made into a windowsill-full by removing whole leaves from the parent plant along with the complete leaf stalk (petiole). The new plant will form at the base of the petiole; so, once you have removed several, dip the bases in rooting hormone and tap off the excess. Insert them into the compost so that the bottom half of the petiole is buried at an angle of 45 degrees.

The new plants will show as clusters of tiny leaves at the bases of the petioles, at which time the plastic can be removed so that the cuttings do not begin to rot. Lift the rooted ones gently and pot them up separately, watering from below, as water on the leaves can cause scorch.

WHOLE LEAF CUTTINGS

Some indoor plants, such as *Begonia rex*, will root from a single leaf laid flat on the compost. Detach a large, healthy, mature leaf, lay it upperside down on a chopping board and use a clean sharp knife to make short cuts through the leaf across the main veins. Lay the leaf vein-side down on fresh, moist compost and weight it in place. The new plants will start to grow as clusters of tiny leaves where the cuts were made and can then be potted separately.

LEAF SECTION CUTTINGS

Certain indoor plants can be grown from small sections of leaf, including *Streptocarpus* cultivars and *Sansevieria* (mother

Leaf cuttings

To ensure that the cuttings are successful:
- *Use fresh, moist seed and cuttings compost*
- *Cover the pot or seed tray with clear plastic to reduce the amount of water being lost from the cuttings. This can be a bag held tightly over a pot with an elastic band, or the whole tray can be placed inside a large bag with the open end closed with a twist tie.*
- *Place the cuttings in a well-lit spot but out of direct sunlight.*
- *Check closely for any sign of infection and remove affected cuttings immediately.*
- *Once you see signs of rooting and new growth, remove and repot the cuttings.*

in law's tongue). The leaves of a streptocarpus can be cut along their length (with the central midrib removed) and placed into the compost, cut-side down, in a shallow trench. The new plants will grow up from the cut edge, and can be potted as soon as they are large enough to handle.

Alternatively, the leaves can be cut across their width and inserted into the compost in an upright position. New plants will form at the bases of the sections.

Increasing plants from cut leaves

The leaves of plants such as Begonia rex *can be used to propagate new plants.*

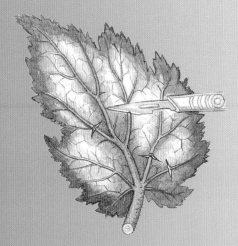

1 Lay the leaf upperside down on a flat surface and make several cuts across the main veins about 1cm (½in) apart.

2 Place the leaf, upperside up, on a tray of compost and hold down the leaf so that the veins are in contact with the compost. Enclose the tray in a plastic bag and keep warm, at 18–24°C (64–75°F). When the plantlets are large enough to handle, pot them up individually.

Left: *Acanthus spinosus* (bear's breeches) does not form a stem, but even the smallest section of root can give rise to several new plants.

Slugs and snails

Root cuttings can be covered with a layer of sharp sand rather than compost to help prevent slugs from eating the new shoots as they emerge from below. Slugs are reluctant to travel over the sharp surface.

root cuttings

If you propagate plants from root cuttings you should bear in mind that they will respond differently at different times of the year. Root cuttings taken in late autumn, when many plants are in the early stages of dormancy, are most likely to be successful, but plants such as alpines are often actively growing in the winter when many other plants are dormant. As with all things, there are exceptions to the rules, and *Armoracia rusticana* (horseradish) seems to grow well whenever it is propagated.

Root cuttings taken in the autumn will usually produce a full-sized plant within 12–18 months.

The preparation and insertion of root cuttings will depend on the type of roots the plant has: thick and fleshy or thin and fibrous.

COLLECTING ROOT CUTTINGS

To collect root cuttings, lift the entire plant and place it at a convenient height on a bench before removing the roots to be used for propagation. This is fairly easy for alpine plants and herbaceous perennials, but digging up an entire tree or climber is clearly not practical. For larger plants a small trench is dug close to the plant to expose a few roots, which can be collected and prepared as cuttings.

After the roots have been collected, the way the plants are treated is determined by the average thickness of the cutting rather than the type of plant from which they are collected.

THICK, FLESHY ROOT CUTTINGS

Plants such as *Acanthus* spp. (bear's breeches) have thick,

Cleaning roots

Before preparing root cuttings, clean off as much soil as possible by washing the roots. This will reduce the amount of damage done to the knife or secateurs used for taking the cuttings.

ROOT CUTTINGS

Plant	Time	Plant type
Ailanthus altissima (tree of heaven)	Winter	Deciduous tree
Aralia spp.	Winter	Deciduous or evergreen tree
Campsis spp.	Winter	Deciduous climber
Celastrus spp.	Winter	Deciduous climber
Clerodendrum spp.	Winter	Deciduous shrub
Rhus spp.	Winter	Deciduous shrub
Robinia spp.	Winter	Deciduous tree
Romneya spp.	Winter	Deciduous sub-shrub
Schefflera elegantissima	Spring	Houseplant (evergreen shrub)

fleshy roots (usually thicker than a pencil), which can be cut into sections about 5cm (2in) long. Make the cut at the top of the cutting at right angles to the root and the bottom one slanting at about 45 degrees. This helps to ensure that the cuttings are not planted upside down.

These thick root cuttings should be inserted vertically into pots filled loosely with loam-based compost. The diameter of the pot is not as important as the depth, which must be at least 8cm (3in) deep to accommodate the cuttings. The slanting end of the root should always be at the bottom, and the top of each cutting should be covered with compost to stop it drying out.

THIN, FIBROUS ROOT CUTTINGS

Plants such as phlox have thin, fibrous roots, usually only several millimetres in diameter. These roots can be cut into 5cm (2in) long sections, with the cuts at the top and bottom of the cutting being made at right angles to the root.

For this type of cutting it is not important to be able to differentiate between the top and bottom of the cutting. These cuttings are not inserted into pots but are scattered over the surface of a tray or pot filled with loam-based compost. The cuttings will lie roughly horizontally over the compost and should be covered with at least 1cm (½in) of compost to prevent them from drying out. These horizontal cuttings will often produce a new plant at each end of the root, while vertical cuttings will usually produce one new plant at the top.

Taking root cuttings

The size of a root cutting will depend on the temperature of the environment in which it is left to grow. In general, the warmer the environment, the more quickly it will grow, and therefore the smaller it can be.

1 *Lift a healthy plant during the dormant season and wash off the soil. Cut off some of the roots close to the crown with a sharp knife. The parent plant can be returned to its position to grow on.*

2 *Cut off any fibrous lateral roots. Make a right-angled cut where the root was severed from the parent. Cut away the thin root end with a sloping cut. Insert into the ground in a coldframe or into a pot in a cold greenhouse.*

grafting

Not every plant that produces tasty fruit or attractive flowers is capable of forming a strong root system, and some plants have roots that are particularly vulnerable to certain soil-borne diseases. This has led to the development of the technique of grafting, which involves taking the strong root system from a plant with undesirable fruit or flowers (the rootstock) and joining it to the topgrowth of a plant with good fruit or flowers (the scion). The two parts join together and grow as one plant. The point at which they are joined is called the graft union and is usually visible as a swelling on the stem of the finished plant.

There are many different

Right: Most types of apple are propagated by some method of grafting onto a rootstock.

Apical-wedge graft

1 The scion should consist of a strong, healthy, one-year-old shoot. Prepare it for grafting by cutting the tip off the shoot just above a leaf joint or node. At the base of the shoot make two slanting cuts 3–4cm (about 1½in) long, to form a wedge. The shoot will have total length of 10–15cm (4–6in).

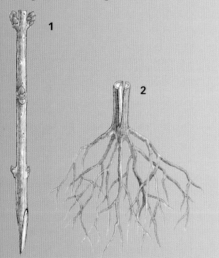

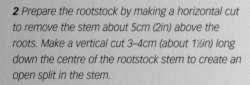
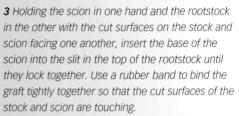

2 Prepare the rootstock by making a horizontal cut to remove the stem about 5cm (2in) above the roots. Make a vertical cut 3–4cm (about 1½in) long down the centre of the rootstock stem to create an open split in the stem.

3 Holding the scion in one hand and the rootstock in the other with the cut surfaces on the stock and scion facing one another, insert the base of the scion into the slit in the top of the rootstock until they lock together. Use a rubber band to bind the graft tightly together so that the cut surfaces of the stock and scion are touching.

4 Place the graft into a pot of compost with the graft union just above the compost surface and water the compost to settle it around the roots. When the graft union has healed, remove the rubber band.

Trees suitable for grafting

Apical wedge
Aesculus, Catalpa, Cercis, Fagus

Side-veneer
Abies, Acer, Betula, Carpinus, Cedrus, Cupressus, Fagus, Fraxinus, Gleditsia, Larix, Magnolia, Picea, Prunus, Robinia, Sorbus

types of graft used for specific plants, but only two basic types of grafting. They are classified according to where the scion is joined onto the rootstock. Thus, if the scion is grafted to the top of the rootstock, it is an apical-wedge graft, and if it is to the side of the rootstock it is an approach graft. Successful grafting relies on there being close contact between the two parts, so they must be positioned carefully and bound closely while they heal together.

Side-veneer graft

For most hardwood trees and deciduous shrubs side-veneer grafting is done in late winter.

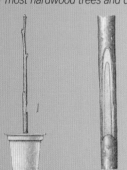 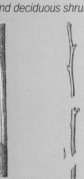 

1 Prepare the rootstock by making a short, downward cut at an angle of about 45 degrees to form a small notch about 2.5cm (1in) above the top root on the rootstock. About 3–4cm (1½in) above this, make a straight downward cut to meet the notch and remove a section of bark and wood from the rootstock.

2 The scion should consist of a strong, healthy, one-year-old shoot, which is prepared for grafting by cutting the tip off the shoot just above a leaf joint or node. Cut a 3–4cm (about 1½in) length of bark from the base of the scion and trim the base to fit snugly into the notch at the bottom of the rootstock cut.

3 Fit the rootstock and scion together, with the cambium (bark) layers of both stock and scion touching one another. Bind the union with a rubber band, which will expand as the union heals and the stem swells. Once the graft has taken, cut back the rootstock at two-week intervals until the top of the rootstock is flush with the top of the graft.

7 BULBS, CORMS AND TUBER

Some of our most familiar garden flowers grow from underground storage organs known as bulbs, corms and tubers. These contain a complete plant (apart from the roots) within their small, roughly circular forms, and all they need is the addition of water and warmth to make them grow.

Left: Dahlias have root tubers, like these *D. 'Gaiety'*. These will not make new plants unless a piece of stem is attached to the tuber.

Bulbs, corms and tubers are underground food-storage organs, but their appearance and method of propagation vary widely. Bulbs (top left) are compressed stems with fleshy scales, which are modified leaves. A corm (top right) is the swollen base of a stem. Tubers (bottom right) are the swollen parts of underground stems, and rhizomes (centre) are horizontally growing stems.

As with other plants, bulbous and cormous plants set seeds, which can be used to produce new stocks, but these plants also reproduce by forming new bulbs and corms on the old ones, and when they are naturalized in the garden they will increase naturally, spreading year by year to create carpets of colour. Some bulbous plants also produce small bulbs on their stems. A crocus corm grows and flowers just once, but as it dies down the energy goes to form one or more new cormlets around the base of the old one. Daffodils produce 'daughter'

bulblets at their bases, as do lilies, but the lily may also produce small bulbils in the leaf joints. Lilies are among the bulbs that can be propagated by scaling, a method of propagation that is almost too fascinating for its own good. It is tempting to repeat it every year to watch the formation of the tiny bulbs on the scales that are removed, especially as the parent bulb will continue to flower. Unless you have unlimited space in your garden, you may find yourself with a great many lilies to find homes for. Although it is possible to remove a large

proportion of the outer scales from a single bulb and it will still flower in the following season, this should not be done year after year because the process will weaken the bulb. It is, however, a useful way of rapidly increasing the numbers of an expensive lily bulb.

Some bulbs are collected in the wild and this practice is leading to the loss of rare species in their natural habitats. When you buy bulbs choose only those that have been raised by nurseries that propagate their own stocks. Better still, try increasing your own stocks.

Left: Snowdrops are often transplanted while they are still in leaf, just after flowering.

Right: Bulbous plants such as daffodils and crocuses can be grown from seeds, but it can take many years before they flower.

bulbs and corms from seeds

Although most bulbs and corms in the garden will be increased by one or other forms of division, many such plants can be propagated by seeds, although hand-pollination may be necessary if several species of the same genus are grown close together. The seeds should usually be used fresh – that is, seeds planted in early autumn of the year in which they ripened will usually germinate, if kept outdoors, in the following spring. Some species need a period of cold temperatures before they will germinate, and some lily seeds take more than a year to germinate.

COLLECTING SEEDS

It is usually recommended that plants such as daffodils and tulips are deadheaded after flowering so that the plants' energies are spent in building up the reserves in the bulb rather than on producing seeds. To collect seeds, however, allow the plants to produce seedheads or capsules, which should be removed as they begin to turn brown. Crocuses and reticulate irises produce seed capsules close to the ground, which can

Bulbs to grow from seeds

When bulbs and corms are grown from seeds it can be between one and five years before the new plants produce flowers. Among the plants that can be propagated from seeds are:

- *Cardiocrinum* spp.
- *Crocus* spp.
- *Freesia* spp.
- *Fritillaria meleagris*
- *Galanthus* spp.
- *Narcissus* spp.
- *Tigrdia pavonia* (peacock flower)
- *Tulipa* spp.

be difficult to see. Remove the seeds from the capsules and keep them in labelled envelopes or paper bags in cool, dry conditions.

SOWING SEEDS

Fill plastic or clay pots with a loam-based compost mixed with fine grit or perlite to make it free-draining. Tamp the compost down lightly and place the seeds evenly over the surface. Sieve a covering of compost over the seeds, aiming to bury the seeds to a depth of four or five times their diameter. Cover this with a layer of coarse grit. Label the pots and plunge them into the ground in a sheltered part of the garden out of direct sun or stand the pots in an open frame. Make sure that the compost does not dry out.

AFTERCARE

When the seedlings have germinated they can be potted up into individual pots or trays in the late summer or early autumn of their first year. The pots should again be sunk into the ground or placed in an open frame. Grow the plantlets on for two or more years until they are ready to be planted out.

Growing bulbs from seeds

Although it will be several years before the bulbs produce flowers, growing from seeds is an inexpensive way to produce large numbers of plants.

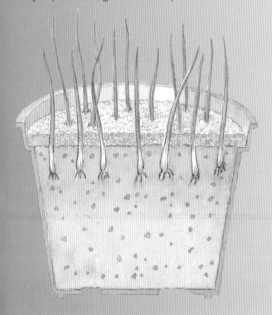

1 *Sow the seeds the autumn after they were collected in a free-draining compost. Cover with sifted compost and cover this with a layer of fine grit so that the seeds are not disturbed. Place the pots in the ground or in an open frame.*

2 *In late summer of their second year the young plants can be potted up into individual pots, where they should be left to grow on for about two years.*

Left: Tulips often produce small bulblets around the base of the parent plant. These bulblets will produce flowers identical to those of its parent bulb.

dividing bulbs

Many bulbs naturally produce offsets or bulblets, and these can be easily separated from the parent plant and replanted. Other bulbs, especially some species of *Lilium* and *Tritonia*, produce bulbils in the leaf axils and bulblets on the underground sections of the stem.

OFFSETS
Some bulbous plants, including narcissi, hyacinths and bulbous

irises, and some cormous plants, such as crocuses and colchicums, produce smaller bulbs and corms around the outer edge of the large one, known as daughter bulbs, bulblets or offsets. These take several years to develop, depending on the species, and are visible only when the parent bulb is dug up. They can easily be detached or left to grow where they are. They should not be separated from the parent until they are old enough to grow on their own, at which point you can gently pull the new bulb away from the old one and pot it separately. It will take up to five years to flower.

BULBILS AND BULBLETS
Plants such as lilies can produce tiny versions of the parent bulb, known as bulbils, above ground in the leaf axils on the stem (the angle where the leaf is attached to the stem). They are small and dark purplish-brown with the same overlapping scales as the parent. As they develop, short

white roots emerge from each bulbil, even while it is still attached to the stem. Once it is ready to grow on its own, it will

Blind bulbs

When permanent plantings of bulbs are blind – that is, they produce leaves but no flowers – the usual cause is overcrowding, although it may also be the result of inadequate water during the growing season. Lift the clump and divide the bulbs, replanting at the recommended depth and at the appropriate planting time.

Plants that produce bulblets

The following species produce bulblets:
- *Allium* spp.
- *Cardiocrinum* spp.
- *Chionodoxa* spp.
- *Erythronium* spp.
- *Fritillaria* spp.
- *Lilium auratum, L. bulbiferum, L.canadense, L. lancifolium, L. longiflorum, L. speciosum*
- *Muscari* spp. (grape hyacinth)
- *Narcissus* spp.
- *Nerine* spp.
- *Scilla* spp.
- *Sternbergia* spp. (autumn daffodil)
- *Tulipa* spp.

Some bulbs, including narcissi, some alliums and crinums, produce 'daughter' bulbs, which can be removed and potted up in spring.

Separating offsets

When clumps of plants such as crocuses and daffodils become overcrowded they should be lifted during their dormancy.

1 Remove as much soil as you can from the bulbs and then carefully pull away the offset bulbs. Rub away the dry skins from the parent bulb, which may be replanted. Large, plump offsets can also be planted directly in the garden.

2 Smaller offsets should be grown on for a year or two. Plant them in free-draining compost, making sure they are at least twice their depth in the pot.

fall from the plant to the soil below and grow as soon as the roots reach moisture.

You will find bulbils around the base of the parent plant and they can be transplanted into pots where you can grow them until they are large enough to be planted out elsewhere. Alternatively, you can gently remove them from the plant as soon as they start to produce their own roots and plant them in pots or seed trays. For the first year the foliage will be grass-like and completely unlike the parent, but it will change in the second year. It will take up to five years before the new plant is ready to flower, but it is an easy means of bulking up your stock. Bulblets are produced by some lilies on the underground sections of the stem, and these can be easily collected if the parent plant is lifted after flowering.

Growing on bulbils and bulblets

Bulbils can be collected from the leaf axils as they ripen. To collect bulblets you will have to lift the parent bulb after flowering.

1 When the leaves have turned yellow, cut the stem from the plant, just above ground level. Gently remove the bulbils from the stem. Place the bulbils on the surface of the compost (usually about ten in a 12–13cm (5in) diameter pot) and gently push them down into the compost until they are just below the surface. Leave them to grow in the pot for a year. The following autumn, shake the bulbs out of the compost and plant them out into the border.

2 Lift the parent bulb and leave it on newspaper for several days for the soil to dry. Shake it gently to remove the dry soil and any loose bulblets from the base of the parent bulb. Place the bulblets on the surface of a pot filled with compost, and gently push them down until they are just below the compost surface. Leave them to grow for a year. The following autumn they can be planted out into the border.

scaling bulbs

Scaling is a procedure widely used on plants such as lilies that is far easier than it sounds and is fascinating to watch. It is a good technique for children to try, and it produces results quickly enough to be interesting. It is also a means of producing many plants from a single bulb, which makes it economical, especially if you have a large area to fill. The best part about it is that it does not affect the flowering ability of the bulb for the coming year.

TIMING
Scaling is done during the autumn period after the leaves die down or when the bulbs on sale in the shops are fresh. If the whole bulb has begun to shrivel it will be of no use for this type of propagation.

CHOOSING THE RIGHT BULB
Choose a bulb that looks healthy and firm and is not showing signs of pest or disease attack – no holes, bite marks, squashy patches or mould. This method of propagation can be used for any of the true lilies, as these all have the same type of bulb. The bulb itself consists of many separate scales, laid in an overlapping pattern around a central point on a hard flat base (the basal plate). The outer scales are often withered and brown but the others should be fleshy and firm.

REMOVING THE SCALES
Pull off the brown outer scales and discard them. The scales you use should be firm and plump to the touch. Bend them back, away from the centre of the bulb, until they snap free. Each should have a piece of the whitish bulb base at the bottom. Take as many as you need, laying them on a clean surface as you pull them off. Up to 80 per cent of the scales can be removed from the bulb without affecting flowering for the following year.

PREPARING THE SCALES
Half-fill a clear plastic bag with fresh, slightly moist, seed and cuttings or multi-purpose compost. Put the scales into the bag until the ratio of compost to scales is about 4:1. Shake the bag gently, so that each scale is in good contact with the compost around it. Fold over the top of the bag and put it in a warm, dark place such as the airing cupboard.

After about two weeks, begin checking the bag regularly for signs of growth and to make sure the compost does not dry out. Try not to open it, but look through the side at the scales. You will see tiny white specks start to appear along the bases of the scales, resembling small grains of rice and known as bulbils. Soon, these start to send out long roots and eventually a single blade-like leaf.

TRANSPLANTING THE SCALES
As soon as a little white bulbil has formed a root, it can be transplanted into its own pot of compost. If a single scale has produced more than one bulbil, you can carefully cut the scale lengthways with a knife so that each bulbil has a piece of the scale still attached (for food) as you plant it.

Once the root adapts to the compost and begins to grow, the scale has served its purpose and will wither and die away. The bulbil has become a plant in its own right, and you will have a whole batch of new lily plants. During their first year they will produce blade-shaped leaves that do not resemble lily leaves at all, but in the following year they will become small versions of the original plant.

It can take up to five years for each plant to become mature enough to flower. Until then, they will need repotting regularly into larger pots to keep them growing well. From about their third year, they can be planted outside, but they will need protection from cold and from pests such as lily beetle until they are established.

When you are planting and transplanting lily scales be careful that you do not plant them too deeply, especially young bulbs, or the bulbs may become blind – that is, they produce plenty of leaves from the base but no stems or flowers – and in some cases the whole bulb may die and rot away.

Scaling lily bulbs

The most important aspect of this method of propagation is to make sure that a small segment of the basal plate (modified stem) is attached to each scale. If there is no basal plate present, or the scale has been broken in half, no embryo plants are present, so no new plants can be produced.

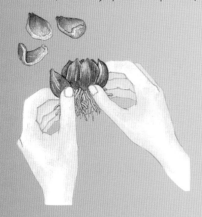

1 The best time to propagate bulbs by scaling is in late summer or early autumn, after the leaves have died down. Choose only vigorous, healthy bulbs with no obvious signs of pest and disease. Remove the outer scales from the bulb, levering them outwards. Work around the bulb removing as many scales as required.

2 Once enough scales have been collected, place them into a clear plastic bag with four times their volume of moist peat and mix the scales thoroughly with the peat. Blow up the bag and close it loosely. Remember to label it on the outside with the date and name of the lily. Place it in a dark, warm place and check it at weekly intervals. Remove any scales that start to decay.

3 After six to ten weeks, two or three small white bulbs will have formed along the bottom edges of the scales, with two or three roots emerging from the base of each miniature bulb. Remove the scales from the compost and pot each scale into a small pot of compost with the miniature bulbs just below the surface (it does not matter if the tip of the scale is visible). Place the pots in a coldframe over winter.

4 For the first season, the bulb leaves will be long and narrow (like grass). By the end of the first growing season, most of the scales will have rotted away, although some will remain more or less intact. Remove the small bulbs from the scales and divide them up individually. These can be planted out into the garden soil or grown in pots for a further year.

Scoring, scooping and chipping

Other techniques used to increase stocks of bulbs are scoring, scooping and chipping. Scoring and scooping are suitable for bulbs such as hyacinths that produce offsets very slowly. Chipping is used for bulbs such as *Hippeastrum*, which do not set seeds and which are difficult to divide. The process of chipping encourages it to produce bulblets.

SCORING
Make two scores at right angles to each other across the basal plate of the bulb. Dust the cut surfaces with fungicidal powder. Place the bulb, upside down, on a tray of sand and put the tray in a warm, dark place until bulblets appear; these can be treated as for bulblets produced by scooping (see below).

SCOOPING
This process is similar to scoring. In late summer remove the basal plate of a bulb with a teaspoon, taking care that you do not damage the outer rim of the basal plate. The aim is to remove only as much tissue as is necessary to reveal the very bottom of the scale leaves. Dust the cut surfaces with a fungicidal powder and set the bulb, upside down, on a tray of sand. Store it in a warm, dry place – an airing cupboard, for example – until bulblets appear on the cut surfaces. Plant the bulb, still upside down, so that the bulblets are just below the surface of the compost. Lift and separate the bulblets at the end of the season and plant them out or pot them up. They should produce flowers after three or four years.

CHIPPING
Cutting the bulbs of plants that are naturally slow to produce

Chipping

Although chipping is not often tried, partly because it is difficult to prevent disease affecting the chipped sections, it is a useful way of propagating bulbs that are slow to produce offsets.

1 Remove the roots and nose with a sharp knife.

2 Cut each bulb lengthways into equal pieces, making sure that each has a section of the basal plate attached to it. Soak in liquid fungicide to protect the chips.

3 Place the chipped sections in moist vermiculite in a plastic bag. Seal the bag and put it in a warm, dark place.

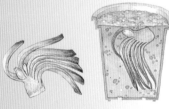

4 After about 12 weeks young bulblets will have begun to appear, and these can be potted up when they have developed roots and are large enough to handle.

Scooping

Cultivars of hyacinths can be propagated by scooping. Although it is possible to buy special tools, a teaspoon with a sharpened edge will be perfectly adequate.

1 With the bulb upside down, scoop out the central part of the basal plate to expose the bottom of the scale leaves. Dust with fungicide.

2 Place the bulb in a tray of sand with the basal plate uppermost. Put the tray in a warm, dark place for about a week, when it will be necessary to moisten the sand to stop the bulbs drying out.

3 After 12–14 weeks bulblets will be visible on the cut surface.

4 Plant the bulb upside down in compost so that the bulblets are just below the surface. In spring leaves will appear and these will die down in late summer, when the bulb can be lifted. Separate the bulblets and pot them up.

offsets stimulates them to produce bulblets. It is essential to use only sterilized equipment throughout the process and to use only well-grown, healthy bulbs. Towards the end of the plant's dormancy cut off the roots just below the basal plate and the nose with a sterilized knife, and then cut the bulb into up to 16 sections, making sure that each

has part of the basal plate. Put some moist vermiculite in a plastic bag and add the bulb sections. Seal the bag and put it in a warm, dark place for about 12 weeks, when bulblets will have developed from the basal plate between the chips. When they have started to form roots and are large enough to handle the bulblets can be potted up individually.

Scoring and scooping

Bulbs that can be propagated by scoring or scooping include:
- *Galanthus* spp. (snowdrop)
- *Hyacinthus* cultivars
- *Lachenalia* spp. (Cape cowslip)
- *Leucojum* spp. (snowflake)
- *Muscari* spp. (grape hyacinth)
- *Narcissus* spp. and cultivars
- *Scilla* spp.

Left: The corms of *Cyclamen neapolitanum* can be cut into segments, but each segment must have an 'eye' or growth bud.

corms and tubers

Many of the most popular garden plants grow from corms and tubers. Corms look similar to bulbs but they are actually modified and swollen stems, which act as food stores. They can be propagated by division. Tubers are swollen stems or roots, which are used for food storage. Among the plants that are grown from tubers are dahlias and some types of begonia and *Tropaeolum*.

Cyclamen and begonias

Some plants present problems when it comes to their propagation, not necessarily because they are difficult to root, but more because there is little or nothing to propagate from. Some types of begonia and cyclamen produce a rounded, swollen stem, which is compressed and may be difficult (if not impossible) to extend sufficiently to produce cuttings from. This means that a slight change of direction is needed to be able to propagate this type of plant successfully. Corms like these can be propagated by division, which involves cutting through the entire corm, rather than a clump.

DIVIDING CORMS

Start by cutting the corm across the middle, cutting it in half and then in half again. Remove and discard any sections which show signs of severe discoloration or rotting. Once the corm has been divided into segments, treat the cut surfaces by dipping them in a shallow tray of sulphur powder. This will encourage the cut surfaces to callus over and reduce the risk of fungal rots developing on the cut surfaces. After the sulphur powder has dried on the cut surfaces, the sections can be planted into pots of compost with the original surface of the corm just above the surface of the compost.

The most important aspect of quartering corms is that, each time the corm is cut, it must contain a piece of the centre segment. It is only the centre of the corm that has the capacity to produce the new roots, leaves, shoots and flowers.

CORMELS

Some corms, such as freesias and gladioli, will produce miniature corms between the union of the old and new corm in the autumn. These can be separated from the corms when they are lifted, and transplanted outdoors the following spring, but it will take at least two or three years of growth before these new corms produce any flowers.

The parent corm of plants such as gladioli will produce many cormels, which can be separated from the parent plant and potted in trays. It will be two or three years before they will produce flowers.

DIVIDING TUBERS

Tubers can also be divided by cutting them into sections – the big difference being that the reproductive parts of a tuber are not always restricted to one area (the potato being a prime example of this).

When you are dividing dahlia tubers make sure that a section of stem is attached to each tuber so that new plants can be produced. Select a healthy tuber and cut it into sections with at least one bud or 'eye' in each section. Discard any sections that show signs of bruising or rotting. After the tuber has been divided into segments, treat the cut surfaces by dusting them with a light covering of sulphur powder. This will encourage the cut surfaces to dry out and reduce the risk of fungal rots developing on the cut surfaces. Place the sections in a seed tray at room temperature until the cut surfaces have fully dried before transplanting them.

Tuberous waterlilies

Waterlily tubers can be divided. Cut off individual sideshoots flush to the tuber. Nymphaea tuberosa develops small protuberances on the tuber, which can be snapped off. Dust the cut surfaces with charcoal and press onto the firmed surface of aquatic compost in a 10cm (4in) pot.

Planting depths

It is important to take great care over replanting the new divisions of corms. If they are planted too deeply, many will fail to grow because they have begun to rot in the wet, airless conditions.

Dividing tubers

Dahlia tubers can be divided to produce flowering plants. Start the tuber into growth before dividing so that you can identify the sections with a viable 'eye'.

1 Place the tubers in a tray of compost until they start into growth. When the 'eyes' are visible, use a sharp knife to separate the sections so that each one has an 'eye'. Dust the cut surfaces with fungicide.

2 Pot up each division into compost and grow on until all danger of frost has passed. Then the plants can be put in their growing positions in the garden.

8 LAYERING

The advantages of layering as a propagation technique are that there is little risk to the plants being propagated and the layers are removed from the parent plant only after they have rooted, which is particularly important for plants that may be propagated in other ways but only if specialized or expensive propagation facilities are available. Plants such as magnolias, rhododendrons and the houseplant *Ficus elastica* (rubber plant), which can be difficult to root from cuttings, can be propagated from layers relatively easily.

Left: Daphnes are notoriously slow to root from cuttings and are ideal for layering.

Right: Some hydrangeas can be propagated by air layering to produce a few large plants.

Below right: Azaleas often have long, arching shoots which are ideal for bending down into the soil.

Home propagation is usually used to replace an existing plant with a younger version of the same plant or to propagate new plants to introduce into the garden. With most forms of layering, only one or two plants can be produced from the existing plant, but they can be relatively large, sometimes up to 1.5m (5ft) in height even before they are separated from the parent plant. Plants of this size quickly give an area of the garden a much more mature appearance, disguising the fact that they are relatively new.

Many plants that are difficult to root from cuttings seem to root easily as layers because the process occurs in almost total darkness. Young, tender stems tend to root more easily and more quickly than older stems because the older tissue is tough and fibrous. Blocking out the light from young stems (by burying them in the soil) will keep them soft and tender, which makes it easier for them to form roots.

A plant that will not root from cuttings can usually be grown from seed, but a disadvantage is the time it can take from seed germination to flowering age. With some plants – wisteria, for example – it can take 12–15 years before a seedling will flower, and the flower can be almost any colour, not necessarily that of its parent. Serpentine layering of wisteria will usually produce a flowering plant within two or three years, and the new plant will carry identical characteristics to its parent.

Left: *Fothergilla major* is very difficult to propagate from cuttings. Layering is the only option for many gardeners.

Below right: *Viburnum opulus* produces long, thin shoots which are very flexible and are ideal for making layers.

types of layering

There are several different methods of layering, and some occur naturally in the garden, when a pliable stem bends down to the ground and develops roots. Eventually a new plant is formed, which will grow independently of the parent plant, even though it may remain attached by the original stem. Commercial nurseries regard layering as a last resort as a method of propagation, partly because it is slow, some plants taking up to two years before they can be severed from the parent plant, and partly because of the small number of new plants that result, but for the gardener this can be an advantage, because you may not want – or may not have room for – as many plants as taking cuttings or sowing seeds can sometimes yield.

The method used will depend on the type of plant:

- Simple layering involves very little more than pegging down a suitable shoot into suitable soil.
- Serpentine and tip layering are similar to simple layering

Constricting stems

The key to successful layering is to convince a section of the plant that it will die if it does not produce roots. This can be done by constricting the section of stem that is covered or is below ground level. Removing a narrow ring of bark, circling the stem with tightly bound wire and splitting or twisting the stem will restrict the flow of food into the end of the layer, forcing it to root.

PLANTS TO LAYER

Plant	Time	Layering method	Time to root
Akebia quinata (chocolate vine)	Early spring to early autumn	Serpentine	12–18 months
Azalea cultivars	Early to late winter	Simple	9–12 months
Buxus spp. (box)	Early to mid-spring	Dropping	6–9 months
Camellia spp.	Late winter to early spring	Simple	9–12 months
Campsis radicans (common trumpet creeper)	Mid- to late autumn	Serpentine	9–12 months
Chimonanthus praecox (wintersweet)	Mid- to late summer	Simple	18–24 months
Clematis spp.	Mid- to late autumn	Serpentine	9–12 months
Daphne spp.	Early to midsummer	Simple	6–9 months
Ficus elastica (rubber plant)	Mid-spring to late summer	Air	6–9 months
Fothergilla spp.	Late summer to early autumn	Simple	12–18 months
Hamamelis spp. (witch hazel)	Early to mid-spring	Air	12–18 months
Jasminum nudiflorum (winter jasmine)	Early to mid-spring	Simple	6–9 months
Magnolia spp.	Early to mid-spring	Simple	12–18 months
Magnolia spp.	Mid-spring to late summer	Air	18–24 months
Parrotia persica (Persian ironwood)	Early to mid-spring	Simple	12–18 months
Rhododendron spp.	Early to mid-spring	Simple	12–18 months
Rubus spp. (blackberry, loganberry)	Early spring to midsummer	Tip	3–6 months
Viburnum spp.	Early to mid-spring	Simple	9–12 months
Vitis spp. (vine)	Early to late winter	Serpentine	9–12 months
Wisteria spp.	Early to mid-spring	Serpentine	12–18 months

in that flexible shoots are buried in the ground.

- Air layering is used for plants like *Ficus elastica* (rubber plant) with upright stems that will not bend down to the ground.
- Dropping is used for dwarf shrubs such as rhododendrons and some dwarf conifers.
- Stooling is a specialized technique that is used for some ornamental shrubs such as *Cornus* spp.

SIMPLE LAYERING

Simple layering is one of the easiest methods of propagation to master because it is 'no-risk'. A shoot is bent down to soil level and pegged down so it is in close contact with the soil, but it is not severed from the parent until it has actually rooted. If the shoot does not root, you can release the shoot to carry on growing normally and try again with another stem. The process is not a quick one, however, so you do need to be patient. With large, established plants, rather than layer one shoot at a time, it is possible to bend down and layer several shoots at once. By doing this you can be pretty sure that at least one of these shoots will form roots at the first attempt.

Left: Akebia quinata (chocolate vine) will root easily as layers made from the non-flowering shoot.

Right: Hamamelis mollis (witch hazel) can be propagated from cuttings and grafting, but layering is still a popular method of increasing plant numbers.

simple layering

Basic layering is a straightforward process – it involves bending down a stem to soil level. It will be some time before the new plant is sufficiently well rooted to be dug up, but by the time you do so it will be a sturdy plant, able to survive the process of being transplanted.

It is a slow method, however, and it can be more than a year before you will be able to separate the new plant and move it to another position. Layering is usually best done in spring, when the sap is rising, but it can also be done between late autumn and early spring.

CHOOSING A SHOOT

The shoot has to be young and flexible enough to be bent over, so look for one of the current year's long, straight growths with no sideshoots. Bend it to test that it is flexible – the last thing you want is to tear off a shoot that is too woody. When you have chosen the shoot select a leaf node about 45cm (18in) from the growing tip. Bend the shoot over and mark the ground where the node will touch it without pulling the shoot away from the plant.

PREPARING THE GROUND

Where you have marked the point where the node touches the soil, prepare the area by clearing away any weeds and large stones. Dig a hole about 15cm (6in) deep; the side nearest the plant should slope at an angle of about 45 degrees, and the side that is furthest from the parent plant should be vertical. Make a U-shaped pin with strong galvanized wire to hold the stem in the hole.

You can root the new plant directly into a pot of compost, rather than the earth, if this is more convenient. Choose a pot that is large enough to accommodate the roots of the layer as they develop, sinking it partially into the soil so that it does not rock or blow over, and peg the stem into it. Once it has rooted you can keep the plant in the container until you are ready to transfer it to its new home.

MAKING THE LAYER

Bend down the stem and lay it in the hole so that the chosen leaf node is at the bottom of the hole. The tip of the shoot will be pushed upright by the vertical side of the hole. Secure the stem in the hole by placing the U-shaped pin directly over the node and pushing it firmly into the soil. Fill in the hole above the stem with loose soil and firm it gently. The upright tip of the shoot can be tied to a short cane to stop it blowing around and loosening the stem from the ground.

To speed up the process and increase the chances of rooting, the stem can be damaged (wounded) at the point where it is pegged into the ground. In the same way that a cutting sends healing hormones to a cut surface, the plant sends help to the damaged section of stem. Use a sharp, clean knife and carefully scrape some bark away from the stem immediately below the chosen leaf node. You can dust the cut with rooting powder for really stubborn subjects.

AFTERCARE

Keep the ground around the layer moist to give it the best chance of rooting. It will take about a year to root properly, but wait until spring before you consider digging it up, which may mean that you leave the plant for up to 18 months. Dig carefully around the new plant to find out if it has rooted (you can often tell by the quantity of growth it has gained). If it has not taken, release the stem and try again.

If it has produced roots, cut the stem leading from the parent plant close to the layer. Dig up the new plant, taking as much rootball as possible, and move it to its new home.

LIFTING AND TRANSPLANTING

One of the hardest parts of layering is deciding when the new

plant has produced enough roots to support it when it is separated from the parent plant, and to complicate matters every plant seems to take a different amount of time to form a new root system.

Usually, the slower-growing the plant, the longer it will take for roots to form on the layer. Most woody plants need at least a year for the layer to develop roots before it can be safely severed from the parent plant. This should be followed by a six-month period when the layer grows independently before it is ready to be lifted and transplanted, which should, ideally, be done during the plant's period of dormancy.

Simple layering

A layer can take a year or more to root, but the resulting plant will be identical to the parent.

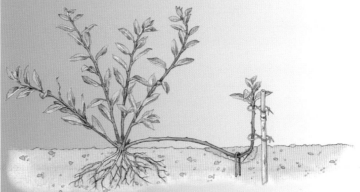

1 *Select a suitable shoot and bend it down until the stem touches the soil at a point at least 30cm (1ft) back from the tip of the shoot. Dig a shallow hole in the ground before lowering the shoot into the bottom of the hole. Hold the stem in the hole with U-shaped pin made of heavy-gauge galvanized wire.*
2 *Replace the soil in the hole around the shoot and firm the area well. Place a vertical cane close to the upright shoot of the layer and tie the shoot to it so that it will grow up straight. Keep the area well watered, especially if the soil is dry. Leave the layer in place until the following spring before you check for rooting.*

Left: Carnations and pinks will layer easily, providing the soil is free-draining.

serpentine layering

This method of layering is similar to simple layering, but it involves trying to create several new plants from a single stem. It is most commonly used for plants with long, flexible, trailing stems, such as clematis, wisteria and rambling roses. The stem is pulled down to ground level and pinned down at intervals along its length.

Serpentine layering is similar to simple layering, but the stem is wounded in several places and pegged down so that there is at least one leaf bud between each wounding site.

Start by wounding the stem that is to be layered. The wounds should be made midway between the leaf nodes (in the area called the internode). At regular intervals, dig shallow holes, 10–15cm (4–6in) deep, in the soil below the shoot, and bend the shoot down so that each wounded area is in the bottom of a hole and the shoot between is still above ground.

Peg the stem into position with U-shaped pins made from heavy-gauge galvanized wire, and fill in the holes with fine soil or compost. Do not layer the tip of the shoot or this will root and the other layered sections behind it will not.

TIP LAYERING

Plants such as blackberries and loganberries can be layered to produce new plants from just the tip of the shoot, where the concentration of rooting hormones is strongest. It is a means of replicating what would happen to the plant if it were growing in the wild. The procedure is basically the same as for simple layering, and the stem can be wounded if it proves to be difficult to root.

In late spring dig a hole, 10–13cm (4–5in) deep, and bend down a strong, healthy shoot so that you can position the tip in the hole. Peg the tip of the shoot in the bottom of the hole with a U-shaped piece of galvanized wire and fill the hole with fine, free-draining soil until the shoot tip is completely covered. Keep the area moist in summer. In autumn dig up the tip of the shoot and check that it has rooted. If it has, separate it from the parent plant; if not, release it to continue growing. The severed section of the layered tip, complete with root system, can be replanted.

Tip layering replicates the process by which *Rubus* spp. propagate naturally. When the tip of the stem is buried it swells and develops roots, eventually forming a new plant.

Layering pinks and border carnations

This is a traditional method of increasing pinks. Before you begin, dig extra peat and sand into the soil around a pink.

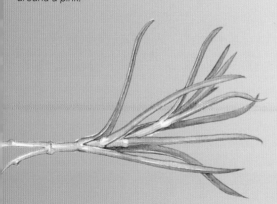

1 *Remove the leaves from the base of a low-growing stem, leaving four or five pairs at the tip. Use a sharp knife to cut the stem to form a tongue extending down from the first leafless node to the next node.*

2 *Pin down the stem into the prepared ground with a U-shaped wire and hold the stem vertical with a cane. Cover the stem with soil. The stem will root after about six weeks and can be severed from the parent for transplanting or potting up.*

Left: Magnolias can be layered into the soil or, if the stems are too stiff to bend, air layering can be used.

air layering

Air layering is a method of propagation used for some indoor plants, such as *Ficus elastica* (rubber plant), and a few outdoor ones, including magnolias. As with ordinary layering the new plant is not removed from the parent until it has rooted, but rather than the stem being bent down to the ground it is treated while it is in its normal growing position. The top of the treated shoot is removed complete with the roots and potted into a container.

The technique is not as complicated as it sounds. It involves coaxing a plant to form roots partway along a stem, as in simple layering, but while the shoot remains upright in its natural growning position rather than being buried in the ground. It is useful for propagating plants with upright shoots that do not bend down easily, such as magnolias and *Hamamelis* spp. (witch hazel), and for indoor plants that are becoming too large for their space, such as *Ficus elastica* (rubber plant). To do this, you will need a knife, a paintbrush, rooting powder, sphagnum moss, a plastic bag, two twist ties and, if wished, a clean matchstick.

Cover the wounded section of the stem with a plastic sleeve filled with sphagnum moss. Eventually the roots will be visible through the plastic.

PREPARING THE STEM

Choose a young shoot that has not become woody; it will be difficult for the layer to root if the stem is too hard. On *Ficus elastica* choose a point about 30cm (12in) back from a growing tip. Wipe clean a section of stem between leaf nodes (especially if it is an outdoor plant) to get rid of any mould or algae.

MAKING THE LAYER

Use a clean knife to make a cut upwards into, but not through, the stem. Cut at an angle of 45 degrees about one-third of the way through the stem, taking care to support the upper part to stop it bending over or breaking. If necessary, tie a cane along the injured part of the stem of thin-stemmed plants to act as a splint and to stop the stem breaking. Carefully bend back the top slightly to open the cut and use a clean paintbrush to apply a little rooting powder to the cut surface. Wedge the cut slightly open with a twist of sphagnum

moss (the type used for hanging baskets) or a clean matchstick.

Take a small, plastic food bag and cut off the bottom, leaving it open at both ends. Slide it carefully over the shoot from the tip down to where you have made the cut in the stem. Use a twist tie to secure it to the stem just below the cut, but not so tightly that the stem is constricted. Pack the bag with the sphagnum moss, so that it is all around the cut stem. Use the second tie to close the bag around the stem at the top, again taking care not to twist it so tight that it bites into the stem.

AFTERCARE

Leave the bag in place and check it regularly. The moss must remain moist around the cut, and you may need to add a little water if it begins to dry out. It will be several weeks before anything happens, and on an outdoor plant it could be six months.

If the layer starts to root you will see roots growing through the moss inside the bag. At this point the layer can be cut from the parent plant just below the new roots. Remove the bag and pot the new plant into fresh compost. It will be vulnerable initially while the roots adapt to growing in compost, so keep even outdoor plants in a sheltered spot for 4–6 weeks.

If no roots appear the layer may not have taken. Remove the bag and the moss so you can check the cut. If the cut is still moist and has swollen callus tissue (the stage before roots appear) around it, replace the packing, using a new bag and fresh moss. If the stem has healed together again, leave it and try again with another shoot.

Dealing with sap

Some plants produce a milky white, sticky latex when they are cut. Spray the cut surface with water several times to wash away the sap, because if it is allowed to dry over the cut surface on the stem it can act as a barrier that will slow down or even prevent the wound healing and forming roots.

Left: Heathers like this
Erica cinerea 'Heathfield'
are often propagated
by dropping.

dropping

Dropping is a layering technique that is often used to propagate many low-growing plants, such as hebes, heathers and dwarf rhododendrons, which do not produce long enough stems to bend down into the soil. These multi-stemmed, shrubby plants tend to become open-centred as they age, with the branches spreading outwards and the centre becoming bare and dead-looking. Rather than pull up the plant and buy another one at a garden centre, you can use the old plant to provide you with several new ones.

PREPARING THE PLANT

Prepare the stock or parent plant in winter by pruning it hard to encourage the development of new, vigorous stems, which should root easily. Some older stems that are left unpruned may respond to dropping, but they never seem to do as well as those that have been pruned.

In spring, before growth begins, dig up the plant, making sure that there is a good ball of soil around the roots. Dig a hole large enough for the whole plant to be 'dropped' into, so that only the tips of the stems are visible 3–5cm (1–2in) above the soil surface. Refill the hole, shaking the soil around the plant's branches and firming it well.

SUMMER CARE

During the growing season the plant must be kept watered enough to encourage roots to start developing. Well-drained soils are better for growing plants for dropping, but they tend to become dry quickly.

REMOVING THE LAYERS

In autumn dig up the entire plant. You should find that most of the branches have formed roots, and most of the new roots will be found on the stems 3–5cm (1–2in) below the soil surface. There may even be some extra branch growth. Cut away each division (young plant) together with a good clump of roots and pot these new plants into containers or plant them out into the garden. Alternatively, if there is room, leave them to grow in a circle around the old plant so that their foliage gradually covers it and you gain an impressive swathe of colour in the garden.

DRAWBACKS

The chief problem with dropping is that the parent plant is either killed by the process or is in such a poor condition after being buried deeply into the soil that it really is not worth the time and effort trying to save it. An additional drawback can be the shape and appearance of the layers that are produced from the parent plant. The type and age of the original plants used for dropping mean that they are often untidy and straggly, with misshapen branches, and, even though the plants are often pruned first, the shape of some branches will not be modified sufficiently to make any significant improvement at all. This may mean that some of the new plants will require considerable remedial pruning in the first and, sometimes, the second year to create desirable shapes.

Dropping

The technique is often used for heathers and dwarf rhododendrons, and it is usually done in spring, when the ground is no longer frozen but before the plant starts into growth.

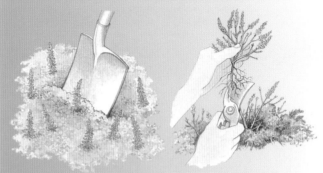

1 *Lift and bury the plant with as much of the rootball as possible, leaving 3–5cm (1–2in) of the tips of the stems exposed above ground.*

2 *In autumn lift the plant and cut each rooted stem away from the parent plant for transplanting or potting up.*

Stooling

The technique can be used to raise plants such as Syringa *spp. (lilac),* Salix *spp. (willow),* Cornus *spp. (dogwood),* Amelanchier *spp. and* Cotinus coggygria. *The parent plant can be used in subsequent years to provide a succession of new plants.*

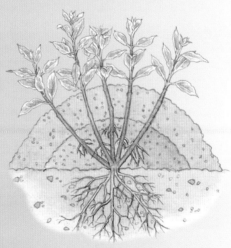

1 *During dormancy cut back all stems to 5cm (2in) from the ground.*

2 *When new shoots develop, mound them up with good-quality soil. Continue earthing them up until about 20cm (8in) of the stem is buried. In early winter carefully remove the mounded soil and sever the rooted stems.*

Left: Plants suffering from severe pest or disease problems like rust should not be used for propagation.

Below right: Aphids not only suck the sap from plants but can also transfer diseases from one plant to another.

troubleshooting

In spite of your best efforts, there will be times when things go wrong with the plants you are trying to propagate or the environment they are in. The art is to minimize the damage, and this will be much easier if you are vigilant and spot a problem before you lose too many plants. Most of the things that go wrong can be dealt with quickly and easily if they are caught in time.

Gardeners tend to blame themselves when something goes wrong with a plant, but there are times when it really is 'one of those things'. Propagation can be a hit-and-miss affair and can depend on factors over which you have no control, such as the weather. Your success rate will increase as you gain experience and learn to understand the plants you are dealing with. Be ruthless with the material you are using and reject any that looks less than perfect. Do not store seeds for too long, because their viability falls with each passing year, and check them regularly to make sure there is no mould in the packets.

One of the commonest problems is fungal attack, because fungi thrive in the conditions of humidity that plants need during their early days. As long as you are using clean equipment and new compost, you are doing all you can to reduce the chances of attack, but spores can enter even the cleanest environment.

Although even clean compost can contain potential diseases, a healthy plant (even a young one) can withstand many of them. They only really start to suffer when they are under another form of stress, and the most common of these is under- or overwatering, both of which affect the way a plant's cells function, leaving it vulnerable. Once an attack starts, the plant is less able to fight it off and can quickly succumb.

Ventilation and temperature also play a part in the spread of a problem. If you are applying extra heat to get the cuttings to root quickly or the seeds to germinate, keep a close watch on the plants. As soon as they have roots, the temperature can be reduced and the ventilation increased. The better the airflow around the plants, the lower the chance of an infection taking hold. As soon as growth starts in earnest, the plants can be moved to a slightly cooler environment to make the conditions less favourable for the disease and help the plant begin to build up its resistance.

The problems that the young plants will face will vary according to where you are doing the propagating, because there will be more pests around if you are working outdoors than inside. Some pests, such as greenflies, are found everywhere, but some, like whiteflies, have more specific requirements and prefer the more humid environment of a greenhouse.

Pests

APHIDS
The small pale green, pink or greeny-black winged or wingless insects may transmit viruses from plant to plant.
Symptoms: Distorted shoot tips and new leaves; sticky coating on leaves, sometimes with black sooty mould.
Prevention: Remove and burn badly infected plants.
Control: Spray at regular intervals with a systemic insecticide, such as bifenthrin, or rape seed oil biological spray as soon as the first aphids are seen in late spring.

CATERPILLARS
Butterfly and moth larvae may have smooth or hairy tubular bodies and dark green or brown heads.
Symptoms: Holes eaten in leaves, flowers and seedpods; the plant may be completely defoliated.
Prevention: Small numbers of plants can be effectively protected by removing the caterpillars by hand.
Control: For large infestations spray plants thoroughly with bifenthrin or the biological control *Bacillus thuringiensis*.

LEAF MINERS
Small insect larvae tunnel around inside leaves as they feed, leaving wiggly lines. They are an aesthetic nuisance rather than harmful.
Symptoms: Pale green or white wiggly lines on leaves.
Prevention: Pull off affected leaves as soon as they are noticed.
Control: Spray with a systemic insecticide, such as bifenthrin or imidacloprid, as soon as the insects are seen.

RED SPIDER MITES
The minute mites suck sap. They can be a serious problem when their populations reach epidemic proportions.
Symptoms: Yellow stunted growth; curled and mottled leaves covered with a fine webbing covering breeding colonies.
Prevention: Spray the undersides of leaves frequently with water and maintain high humidity levels.

Control: Spray with a systemic insecticide, such as bifenthrin or imidacloprid, regularly as soon as noticed, or use the parasitic insect *Phytoseiulus* to control biologically.

SCALE INSECTS
The small, brown, blister-like bumps on stems and leaves suck sap, gradually weakening the whole plant.
Symptoms: Stunted growth and yellowing of leaves; sticky coating on lower leaves, sometimes with black sooty mould.
Prevention: Barrier glue around the stem stops the larval stage moving to new sections of the affected plant.
Control: Introduce the predator *Metaphycus* in midsummer; apply an organic insecticide based on fatty acids or a systemic insecticide spray, such as imidacloprid, in late spring and early summer.

Left: Coral spot frequently starts on dead wood before spreading into live tissue and killing branches and stems.

SCIARID FLIES

Tiny black flies, that are also known as mushroom flies or fungus gnats, are found in loam-free composts. The larvae feed on plant roots.

Symptoms: Stunted growth and wilting plants; seedlings and cuttings that have poor root systems.

Prevention: Avoid overwatering the compost and using poorly draining compost mixes; do not use unsterilized compost.

Control: Treat the compost with imidacloprid granules.

SLUGS AND SNAILS

Slugs have slimy, tubular-shaped bodies, ranging in colour from creamy white to jet black. Snails are similar but have a semicircular shell on their back.

Symptoms: Holes are eaten in plant tissue overnight; the damaged seedlings are usually killed.

Prevention: Apply a sharp mulch of gravel around the plants as a barrier.

Control: Place slug pellets around the base of the plants; position traps made of grapefruit skins or use saucers of beer; water nematodes into warm, moist soil.

VINE WEEVILS

White, legless, C-shaped grubs with black-brown heads feed on roots; black-brown, long-nosed adults eat leaves.

Symptoms: Plants wilt or collapse, as most of the roots are eaten; small semicircular notches are bitten out of the leaf edges.

Prevention: Keep compost surface clear of debris, which hides adult female insects.

Control: Treat plants with imidacloprid spray on the leaves or treat the compost with imidacloprid granules or parasitic nematodes, which are active at soil temperatures of 14°C (57°F) or more.

WHITEFLIES

Tiny, white, moth-like insects cluster on the undersides of the leaves and suck sap.

Symptoms: Often a general decline in plant vigour and possibly symptoms of virus as these insects can pass viruses from plant to plant.

Prevention: Planting *Tagetes* cultivars close by will help to draw the insects away from the plants.

Control: Under glass, introduce the parasitic wasp *Encarsia formosa* to control the young whiteflies.

Diseases

BOTRYTIS (grey mould)

The fungus infects flowers, leaves and stems. It usually enters through wounds in the plant, such as cuts and bruises.

Symptoms: Discoloured, yellowing leaves die slowly; stems may rot at ground level; the entire plant is covered with a grey, felt-like mould.
Prevention: Remove affected plants at once; maintain good air circulation.
Control: Spray with carbendazim as soon as the disease symptoms are seen.

CORAL SPOT

Although this fungus commonly invades dead wood it may also invade live tissue, including cuttings and layers.
Symptoms: Individual branches wilt in summer; grey-brown staining may be found under the bark; in autumn the branch is covered in small salmon-pink blisters.
Prevention: Prune in summer when there are few fungal spores in the air; also clear away old prunings.
Control: Remove and burn infected material as quickly as possible.

DAMPING-OFF

This fungal problem is especially associated with seedlings.
Symptoms: The tops of seedlings shrivel, curl over and die; the problem spreads rapidly to the whole pot or seed tray.
Prevention: Maintain good air circulation around the young plants and avoid excess humidity; do not use unsterilized compost.
Control: Spray with copper sulphate and ammonium carbonate combination.

DOWNY MILDEW

The fungus infects leaves and stems and can also overwinter in compost or plant debris.

Symptoms: Discoloured, yellowing leaves have white patches on underside; plants often die slowly in autumn.
Prevention: Avoid over-crowding, to maintain good air circulaiton, and do not use unsterilized compost.
Control: Spray with mancozeb as soon as the disease symptoms are seen or remove and burn badly infected plants and cuttings.

POWDERY MILDEW

The parasitic fungal disease invades the soft tissue inside leaves and feeds off the plant.
Symptoms: White floury patches appear on young leaves; shoots become distorted; leaf-fall is premature.
Prevention: Prune out infected stems in autumn to prevent spores overwintering.
Control: Spray carbendazim on the young leaves (the fungus can not penetrate old leaves) at the first signs of infection.

ROOT ROTS

Roots are attacked by spores that build up both in the compost and within infected plant remains.
Symptoms: Foliage turns yellow; branches die back from the tips; affected roots usually turn black.
Prevention: Avoid heavy watering and improve drainage; do not use unsterilized compost.
Control: Remove and burn affected plants; select cultivars known to be tolerant or resistant.

RUST

The fungal disease attacks many plants, including roses, weakening growth and leading to premature leaf-fall.

Symptoms: Yellow blotches on the surface of leaves; bright orange or brown patches of spores on undersides.
Prevention: Space plants more widely to improve circulation of air; increase ventilaiton.
Control: Remove all affected areas by hand; affected plants should be thoroughly sprayed regularly with mancozeb.

SILVER LEAF

The fungus enters the woody tissue of members of the *Prunus* cherry family, both ornamental and fruiting.
Symptoms: Leaves of infected trees develop a silvery sheen; branches die back; brownish-purple brackets appear on stems.
Prevention: Prune in summer when there are few fungal spores in the air.
Control: Prune infected branches from otherwise healthy trees; badly infected trees must be removed and burned.

Viruses

Microscopic infections are often carried by sap-feeding pests, such as aphids, and passed from plant to plant.
Symptoms: Leaves (and shoots) are small, distorted or grouped in rosettes; yellow discoloured patterns on leaves.
Prevention: Buy virus-free plants; control potential carriers; clear away weeds that may harbour viruses.
Control: Remove and burn affected plants as soon as possible. Do not propagate from them.

*Clockwise from left: Acer Palmatum
'Lutescens', Papaver somniferum,
Dianthus 'Diane'.*

directory of plants

Some of the plants we see most often in gardens are popular because they are easy to grow and propagate. Some plants, such as forsythia, can be propagated from four different types of cutting, so that they can be propagated easily and all year round.

Rather than make the art of propagation a complicated issue, especially if you are a beginner, it is better to use the simplest techniques and facilities you can successfully get away with.

There is a wide range of plants that only need some space in the garden, so that layers can be bent down to the ground, hardwood cuttings inserted, or divisions replanted. Others, which are slow to root but do not need warm conditions, will often do well in a small coldframe, and those which need a little heat will often do well on the kitchen windowsill.

ANNUALS AND PERENNIALS

Plant	Method	Timing	Place	Time taken
Acanthus spp. (bear's breeches)	Division	Spring or autumn	Garden soil	6–8 months
Achillea spp. (yarrow)	Division	Spring	Garden soil	6–8 months
Agapanthus spp. (African blue lily)	Division	Spring	Garden soil	6–8 months
Alchemilla mollis (lady's mantle)	Division	Spring or autumn	Garden soil	6–8 months
Anemone x *hybrida*	Division	Spring or autumn	Garden soil	6–8 months
Antirrhinum cultivars	Seed	Summer to autumn or spring	Windowsill	2–3 months
Aquilegia spp.	Division	Spring	Garden soil	6–8 months
Aster spp.	Division	Spring	Garden soil	6–8 months
Bassia spp. (burning bush)	Seed	Spring	Windowsill	2–3 months
Begonia spp.	Seed	Winter	Windowsill	2–3 months
Bergenia spp.	Division	Spring or autumn	Garden soil	2–3 months
Calendula spp.	Seed	Spring or autumn	Garden soil	2–3 months
Capsicum annuum	Seed	Winter	Windowsill	2–3 months
Chrysanthemum cultivars	Basal cutting	Winter to spring	Windowsill	2–3 weeks
Clarkia amoena (godetia)	Seed	Autumn or spring	Garden soil	2–3 months
Convallaria majalis (lily-of-the-valley)	Division	Autumn	Garden soil	6–8 months
Cosmos spp.	Seed	Late spring	Garden soil	2–3 months
Delphinium cultivars	Basal cutting	Early spring	Garden soil	6–8 months
Dianthus spp. (pink)	Softwood cutting	Summer	Windowsill	2–3 months
Digitalis purpurea (foxglove)	Seed	Late spring	Garden soil	12 months
Echinacea purpurea (coneflower)	Division	Autumn or spring	Garden soil	6–8 months
Eremurus cultivars (foxtail lily)	Division	Autumn	Garden soil	6–8 months
Eschscholzia californica (California poppy)	Seed	Spring or autumn	Garden soil	2–3 months
Fragaria spp. (strawberry)	Runners	Summer	Garden soil	8–12 months
Fuchsia cultivars	Softwood cutting	Spring	Windowsill	1 month
Geranium spp.	Division	Summer	Garden soil	1 month
Gypsophila spp.	Softwood cutting	Summer	Coldframe	2–3 months
Helleborus spp.	Division	Spring or late summer	Garden soil	12 months
Hemerocallis cultivars (daylily)	Division	Spring or autumn	Garden soil	6–8 months
Hosta cultivars	Division	Late summer	Garden soil	6–8 months
Impatiens balsamita	Seed	Early spring	Windowsill	2–3 months
Iris germanica	Division	Late summer to early autumn	Garden soil	12 months
Lathyrus odoratus (sweet pea)	Seed	Spring	Windowsill	2–3 months
Lobelia erinus cultivars	Seed	Late winter	Windowsill	2–3 months
Matthiola incana	Seed	Early spring	Garden soil	2–3 months
Nicotiana spp.	Seed	Spring	Windowsill	2–3 months
Papaver annual spp.	Seed	Spring	Garden soil	2–3 months
Papaver perennial spp.	Division	Spring	Garden soil	12 months
Pelargonium cultivars	Softwood cutting	Spring, summer or autumn	Windowsill	1 month

Petunia cultivars	Seed	Autumn or spring	Windowsill	2–3 months
Phlox cultivars	Root cuttings	Spring or early autumn	Coldframe	8–12 months
Salvia annual spp.	Seed	Spring	Windowsill	2–3 months
Saxifraga spp.	Offsets	Spring or summer	Garden soil	1–3 months
Schizostylis coccinea	Division	Spring	Garden soil	12 months
Sedum spectabile	Division	Spring	Garden soil	8–12 months
Sempervivum spp.	Offsets	Spring	Garden soil	1–3 months
Tagetes spp.	Seed	Late spring	Garden soil	2–3 months
Tropaeolum cultivars	Seed	Spring	Garden soil	2–3 months
Viola wittrockiana cultivars	Seed	Winter or spring	Garden soil	8–12 months

BULBS, CORMS AND TUBERS

Plant	Method	Timing	Place	Time taken
Allium spp.	Division	Spring	Garden soil	2–3 months
Begonia tuberous cultivars	Division	Spring	Windowsill	3–4 months
Crocosmia cultivars	Division	Spring	Garden soil	12 months
Cyclamen persicum	Seed	Summer	Windowsill	6–8 months
Dahlia cultivars	Division	Spring	Garden soil	6–8 months
Galanthus spp. (snowdrop)	Division	Late spring	Garden soil	12 months
Gladiolus cultivars	Cormlets	Spring	Garden soil	1–3 years
Lilium spp.	Scales	Late summer	Windowsill	3–4 months

GRASSES AND BAMBOOS

Plant	Method	Timing	Place	Time taken
Briza spp.	Division	Spring to summer	Garden soil	4–6 months
Cortaderia spp. (pampas grass)	Division	Spring	Garden soil	6–8 months
Fargesia spp. (bamboo)	Division	Spring	Garden soil	6–8 months
Miscanthus spp.	Division	Spring	Garden soil	4–6 months
Phyllostachys spp.	Division	Spring	Garden soil	6–8 months

CLIMBERS

Plant	Method	Timing	Place	Time taken
Clematis cultivars	Softwood cutting	Spring	Windowsill	6–8 months
Hedera spp. (ivy)	Semi-ripe cutting	Summer	Coldframe	4–6 months
Lonicera spp.	Hardwood cutting	Autumn	Garden soil	12 months
Vitis spp.	Hardwood cutting	Late winter	Coldframe	12 months
Wisteria spp.	Serpentine layering	Autumn	Garden soil	8–12 months

TREES AND SHRUBS

Plant	Method	Timing	Place	Time taken
Acer spp. (maple)	Layering	Autumn	Garden soil	12–18 months
Azalea cultivars	Layering	Autumn	Garden soil	12–18 months
Berberis spp.	Semi-ripe cutting	Summer	Coldframe	18 months
Buddleja spp.	Hardwood cutting	Autumn to spring	Garden soil	12 months
Buxus spp. (box)	Semi-ripe cutting	Summer	Coldframe	18 months
Calluna cultivars (heather)	Semi-ripe cutting	Summer	Coldframe	18 months
Camellia spp.	Semi-ripe cutting	Autumn to winter	Windowsill	6–8 months
Chamaecyparis spp. (cypress)	Semi-ripe cutting	Late summer	Coldframe	8–12 months
Cordyline spp.	Stem cutting	Spring	Windowsill	6–8 months
Cornus spp. (dogwood)	Hardwood cutting	Autumn	Garden soil	12 months
Cotoneaster evergreen spp.	Semi-ripe cutting	Late summer	Coldframe	18 months
Deutzia spp.	Hardwood cutting	Autumn	Garden soil	12 months
Erica spp. (heather)	Semi-ripe cutting	Late summer	Coldframe	8 months
Escallonia cultivars	Semi-ripe cutting	Late summer	Windowsill	18 months
Euonymus evergreen spp.	Semi-ripe cutting	Summer	Coldframe	6–8 months
Fagus spp. (beech)	Seed	Autumn	Garden soil	8–12 months
Forsythia intermedia	Hardwood cutting	Autumn	Garden soil	12 months
Hamamelis spp. (witch hazel)	Layering	Autumn	Garden soil	12–18 months
Jasminum spp.	Layering	Autumn	Garden soil	4–6 months
Mahonia spp.	Semi-ripe cutting	Summer to autumn	Coldframe	8–12 months
Paeonia spp.	Division	Autumn or spring	Garden soil	12 months
Potentilla spp.	Semi-ripe cutting	Early summer	Windowsill	2–3 months
Pyracantha spp.	Semi-ripe cuttings	Summer	Coldframe	8–12 months
Quercus spp. (oak)	Seed	Autumn	Coldframe	8–12 months
Ribes spp. (currant)	Hardwood cutting	Winter	Garden soil	12 months
Taxus spp.	Semi-ripe cutting	Late summer	Coldframe	8–12 months
Thuja spp.	Semi-ripe cutting	Late summer	Coldframe	8–12 months
Weigela spp.	Hardwood cutting	Autumn to winter	Garden soil	12 months

HOUSEPLANTS

Plant	Method	Timing	Place	Time taken
Billbergia spp.	Division	Spring or summer	Windowsill	2–3 months
Chlorophytum comosum	Division	Spring, summer or autumn	Windowsill	2–3 weeks
Ficus elastica	Air layering	Spring or summer	Windowsill	8–12 months
Kalanchoe spp.	Stem cuttings	Spring or summer	Windowsill	2–3 months

CALENDAR

The season indicated for propagating each plant is a guide only, and will depend on the individual species or cultivar and on the weather and conditions in your garden. Many bulbs, for example, are divided after flowering, which may, for spring-flowering cultivars, be any time from early spring to early summer. Herbaceous perennials are often divided in autumn, but, again, this will depend on the plant, the weather and, in some instances, your own preferences.

Spring

Seeds

Hardy annuals and biennials
Bellis perennis, Calendula officinalis, Clarkia amoena (godetia), *Eschscholzia californica, Gypsophila elegans, Helianthus annuus, Helichrysum bracteatum, Lathyrus odoratus Lobularia maritima, Matthiola* spp., *Myosotis* spp., *Nigella damascena, Papaver rhoeas, P. somniferum, Primula* (polyanthus) cvs., *Tropaeolum majus*

Bedding plants and half-hardy annuals
Ageratum houstonianum cvs., *Amaranthus* spp., *Antirrhinum majus, Bassia scoparia, Cosmos* spp., *Dianthus* (annual carnation) cvs., *Dorotheanthus bellidiformis, Impatiens walleriana* cvs., *Lobelia erinus* cvs., *Nemesia strumosa, Nicotiana* spp., *Pelargonium* spp. and zonal cvs., *Petunia* cvs., *Salvia splendens, Senecio cineraria, Tagetes erecta* (African marigold), *T. patula* (French marigold), *Verbena* x *hybrida* cvs.

Perennials
Agapanthus spp., *Delphinium* cvs., *Erysimum* cvs., *Verbascum* spp.

Houseplants
Asparagus setaceus, A. densiflorus Sprengeri Group, *Begonia scharffii, Capsicum annuum, Campanula isophylla, Hypoestes phyllostachya, Impatiens* spp., *Solenostemon* cvs.

Vegetables
Aubergine, celery, courgette, cucumber, melon, pepper, sweetcorn, tomato

Division

Herbaceous perennials
Achillea spp., *Alchemilla mollis, Aster* cvs., *Astilbe* cvs., *Bergenia* spp., *Cortaderia selloana, Delphinium* cvs., *Dicentra* spp., *Eryngium* spp., *Helleborus* spp., *Hosta* cvs., *Hypericum* spp., *Kniphofia* cvs., *Ligularia dentata, Monarda* spp., *Ophiopogon* spp., *Paeonia* cvs., *Sedum* spp., *Tradescantia - andersoniana* cvs.

Bulbs, corms and tubers
Begonia tuberhybrida cvs., *Chionodoxa* spp., *Dahlia* cvs., *Galanthus* spp., *Hyacinthus* cvs., *Tulipa* cvs.

Shrubs
Penstemon spp., *Potentilla* spp. *Syringa.*

Layering

Trees, shrubs and climbers
Buxus, Hamamelis spp., *Jasminum* spp., *Magnolia* spp., *Parrotia persica, Prunus* spp., *Rhododendron* cvs., *Syringa, Tilia* spp., *Viburnum* spp., *Wisteria* spp.

Cuttings
Trees, shrubs and climbers
Callicarpa spp., *Campsis radicans*, *Catalpa* spp., *Clematis* spp., *Forsythia x intermedia*, *Fuchsia* cvs., *Hedera* spp., *Kolkwitzia amabilis*, *Mahonia* spp., *Perovskia atriplicifolia*, *Romneya coulteri*, *Vitis* spp.

Houseplants
Begonia spp., *Campanula isophylla*, *Codiaeum* cvs., *Crassula* spp., *Dieffenbachia seguine*, *Ficus* spp., *Hoya carnosa*, *Impatiens* spp., *Kalanchoe* spp., *Monstera deliciosa*, *Saintpaulia* cvs., *Schleffera elegantissima*, *Stephanotis floribunda*, *Tradescantia fluminensis*.

Summer
Seeds
Hardy annuals and biennials
Digitalis spp., *Dipsacus fullonum*, *Lunaria annua*, *Oenothera biennis*, *Viola x wittrockiana* cvs.

Houseplant
Cyclamen persicum.

Division
Herbaceous perennials
Agapanthus africanus, *Iris germanica*.

Bulbs, corms and tubers
Fritillaria spp., *Narcissus* cvs.

Layering
Trees, shrubs and climbers
Akebia quinata, *Chimonanthus praecox*, *Daphne* spp., *Rubus* spp.

Cuttings
Trees, shrubs and climbers
Abelia spp., *Actinidia kolomikta*, *Aucuba japonica*, *Azara* spp.,
Buddleja spp., *Buxus* spp., *Callicarpa* spp., *Camellia* spp., *Campsis radicans*, *Catalpa* spp., *Ceanothus* spp., *Celastrus* spp., *Chaenomeles* spp., *Cistus* spp., *Clematis* spp., *Cornus* spp., *Cotinus* spp., *Cytisus* spp., *Deutzia* spp., *Elaeagnus* spp., *Escallonia* spp., *Fallopia baldschuanica*, *Fatsia japonica*, *Forsythia x intermedia*, *Fuchsia* cvs., *Garrya elliptica*, *Hebe* spp., *Hydrangea* deciduous spp., *Hypericum* spp., *Kerria japonica*, *Kolkwitzia amabilis*, *Lonicera* spp., *Magnolia* spp., *Osmanthus* spp., *Parthenocissus* spp., *Perovskia atriplicifolia*, *Philadelphus* spp., *Pittosporum* spp., *Potentilla* spp., *Pyracantha* spp., *Rhododendron* cvs., *Ribes* evergreen spp., *Rubus* spp., *Schisandra* spp., *Schizophragma hydrangoides*, *Solanum* spp., *Spiraea* spp., *Stewartia* spp., *Syringa* spp., *Viburnum* spp., *Wisteria* spp.

Houseplants
Aphelandra spp., *Justicia* spp.

Autumn
Seeds
Vegetables
Broad beans, garlic

Division
Herbaceous perennials
Anemone spp., *Echinops* spp., *Erigeron* spp., *Geranium* spp., *Helleborus* spp., *Hemerocallis* cvs., *Phlox* spp., *Rudbeckia* spp.

Bulbs, corms and tubers
Allium spp., *Camassia* spp., *Crocosmia* spp., *Crocus* spp., *Convallaria majalis*, *Gladiolus* spp., *Lilium* spp., *Muscari* spp., *Cornus* spp., *Fargesia* spp., *Kerria japonica*.

Layering
Trees, shrubs and climbers
Campsis spp., *Clematis* spp., *Fothergilla* spp.

Cuttings
Trees, shrubs and climbers
Berberis evergreen spp., *Brachyglottis* spp., *Buddleja davidii*, *Chamaecyparis* spp., *Choisya ternata*, *Cupressocyparis*, *Cupressus* spp., *Deutzia* spp., *Euonymus* spp., *Forsythia x intermedia*, *Ilex* spp., *Itea virginica*, *Juniperus* spp., *Metasequoia glyptostroboides*, *Olearia* spp., *Phlomis* spp., *Potentilla fruticosa* cvs., *Prunus lusitanicus*, *Pyracantha* spp., *Rosa* cvs., *Taxodium* spp., *Taxus baccata*, *Thuja* spp., *Weigela*.

Perennials
Fuchsia cvs., *Pelargonium* cvs.

Winter
Layering
Trees, shrubs and climbers
Acer spp., *Camellia* spp., *Hydrangea* spp., *Rhododendron* (azalea) cvs., *Vitis* spp.

Cutting
Trees, shrubs and climbers
Ailanthus altissima, *Aralia* spp., *Buddleja* spp., *Campsis* spp., *Clematis* spp., *Clerodendron* spp., *Cornus* spp., *Leycesteria formosa*, *Ligustrum ovalifolium*, *L. vulgare*, *Platanus* spp., *Populus* spp., *Rhus typhina*, *Ribes* deciduous spp., *Salix* spp., *Sambucus* spp., *Symphoricarpus* spp., *Vitis* spp., *Weigela* spp., *Wisteria* spp.

Houseplants
Cacti.

Acclimatize (acclimate) To accustom plants to conditions that are different (usually cooler) from those in which they are growing. See also HARDEN OFF.

Adventitious A bud or root that arises in an unusual or unplanned place.

Alkaline A substance, such as soil, with a pH value of more than 7·0.

Alpine A plant originating from mountainous regions, growing between the tree and snow lines; often applied loosely to a wide range of rock garden plants.

Alternate Branches or leaves that occur at different levels on opposite sides of the stem.

Annual A plant that completes its reproduction cycle (grows from seed, flowers, sets seeds and dies) in one year.

Apical bud The uppermost bud in the growing point of a stem; also known as the terminal bud.

Auxin A naturally occurring hormone that controls the growth of a plant.

Axil The angle formed between the main stem of a plant and a leaf stalk growing from it.

Axillary bud A bud that occurs in a leaf axil.

Bark The protective layer of dead cells on the outer surface of plant roots and the stems of woody plants.

Basal A shoot or bud arising from the base of a plant.

Basal plate The compressed stem of a bulb.

Bedding plant An annual or biennial raised under glass before being planted out in beds or borders to create a temporary display.

Biennial A plant that completes its growing cycle in two growing seasons. It germinates and produces roots and leaves in its first year, and flowers and produces seed before dying in the second year.

Blind A plant that fails to produce flowers or a stem with a damaged growing tip.

Bract A modified leaf, often brightly coloured (as on clematis and poinsettias) and appearing flowerlike.

Branch A shoot growing directly from the main stem of a woody plant.

Broadcast To spread fertilizer or seeds randomly onto prepared ground but without pre-drilling holes or furrows.

Bud A condensed shoot containing an embryonic shoot or flower.

Budding A form of grafting (often used on roses and fruit trees), by which two plants are joined together.

Bud union The point at which a scion is budded on to a rootstock.

Bulb A storage organ, consisting of thick, fleshy scale leaves arranged on a compressed stem (basal plate), that is found below soil level.

Bulbil A small bulb or bulb-like growth found in the leaf axil of an upright stem.

Bulblet A small bulb that develops on the side of the parent bulb.

Callus The soft plant tissue that forms as a protective cover over a cut or wounded surface.

Cambium layer The layer of living cells that lie immediately under the bark and are responsible for stem thickening and healing.

Capping The crust that forms on the surface of soil.

Chilling A period of low temperature required by plants during their dormancy to stimulate flower development.

Chip (nick) To damage the hard outer case of a seed to encourage germination; also known as SCARIFY.

Clone A collective term for a number of identical plants that have been propagated by cuttings or division from a single individual.

Coldframe A low, portable or permanent glazed structure used for protecting plants.

Collar The point at the base of the main stem of a plant where the roots begin.

Compost A potting medium (soil mix) made to a standard formula. The mixture may contain loam, sand, leaf mould or peat (or peat substitute).

Corm An underground modified stem forming a storage organ.

Cormlet A small corm formed around the base of the parent corm.

Cotyledons The seed (first) leaves on a seedling. Plants may be dicotyledonous (with two seed leaves) or monocotyledonous (with one seed leaf).

Crown The top part of a rootstock from which shoots arise.

Cultivar A cultivated variety rather than a naturally occurring hybrid.

Cutting A section of a plant's stem, root or leaf that is used for propagation.

Dioecious A plant (such as willow or holly) that has male and female flowers on separate plants.

Division A method of propagation used to increase the number of plants by splitting them into smaller units, each of which has a root system and one or more shoots.

Dormancy A period when a plant is resting, usually in winter, or a period of suspended growth.

Drill A narrow, straight, shallow furrow into which seeds are sown.

F1 hybrid First-generation seed produced by crossing two pure-bred strains. An F2 hybrid is a second-generation cross between two F1 hybrids. The second generation will rarely breed true.

Force To induce plants to start growing earlier than they usually would.

Germination The first stage in the development of a seed into a plant.

Grafting A propagation method involving the artificial joining of two or more separate plants.

Graft union The point at which a scion is grafted onto a rootstock.

Harden off To accustom plants to conditions that are different (usually cooler) from those in which they are growing. See also ACCLIMATIZE.

Heel The piece of bark attached when a shoot is pulled, rather than being cut, from the parent plant.

Heel in To put a plant in a temporary position until it can be planted out in its final position.

Herbaceous A non-woody plant with annual topgrowth and a perennial root system or storage organ.

Lateral A stem or shoot arising from an axillary bud.

Layering A method of propagation by which roots form on a stem before it is detached from the parent plant.

Leader The dominant shoot or stem of the plant (usually the terminal shoot).

Loam A fertile soil containing equal proportions of clay, sand and silt.

Maiden A young (one-year-old) budded or grafted tree or bush.

Meristem Plant tissue that possesses the capacity to divide and multiply (found in buds, roots and stems).

Mulch A layer of material, which may be organic or inorganic, applied to cover the soil, to suppress weeds and conserve moisture.

Mutation A plant change or variation occurring by chance, often referred to as a sport.

Opposite Leaves, buds or stems that are arranged on shoots in pairs directly opposite one another.

Perennial Any plant that lives for three or more years.

pH The measure of acidity and alkalinity in a soil. A pH of 7 indicates neutral soil; soils with a pH lower than 7 are acid, and those with a pH higher than 7 are alkaline.

Pinch out To remove (usually with the finger and thumb) the growing point of a shoot to encourage the development of laterals or sideshoots. See also STOP.

Pot on To transfer a plant to a larger container.

Pot up To transfer young plants or seedlings from the seedbed or seed tray to a larger container.

Prick out To transfer seedlings from a seedbed to their final growing position.

Propagator A glass or plastic structure, which may be heated, in which seedlings and cuttings are raised.

Rhizome A specialized underground stem (of plants such as *Begonia rex* and *Convallaria majalis*), which lies horizontally in the soil and produces roots and shoots.

Root prune To cut back live roots to control a plant's vigour.

Rootstock The crown and root system of plants such as herbaceous perennials. The plant onto which another plant, the SCION, is budded or grafted.

Runner A stem bearing a new plant that grows horizontally along the ground (as with strawberries).

Sap The juice or fluid in a plant.

Scale The modified leaf (fleshy segment) of a bulb, such as an onion or hyacinth.

Scarify To damage the hard coating of a seed, by nicking, filing or otherwise abrading, in order to encourage germination; also known as CHIP.

Scion The bud, shoot or cutting that is used for budding or grafting onto a ROOTSTOCK.

Spur A short fruit- or flower-bearing branch.

Stool To cut down plants to within 10–15cm (4–6in) of ground level to encourage new growth.

Stop To cut out the growing point of a shoot to encourage the development of lateral shoots. See also PINCH OUT.

Stratify To store seed in cold or warm conditions to overcome dormancy.

Sub-lateral A sideshoot arising from an axillary bud of a lateral shoot.

Sucker A shoot arising from a plant's roots or from an underground stem.

Taproot The main, anchoring root of a plant.

Terminal bud The uppermost bud in the growing point of a stem; also known as the apical bud.

Tip prune To cut back the growing point of a shoot to encourage the development of lateral shoots.

Transplant To move plants from one growing area to another in order to provide them with more growing room.

Tuber A fleshy root (as of a dahlia) or stem (as of a potato) modified to form a storage organ.

Union (graft union) The point at which a scion is grafted onto a rootstock.

Whip A one-year-old tree with no lateral (side) branches.

INDEX

ACKNOWLEDGEMENTS

Garden Picture Library/MIchael Howes 52, /Jacqui Hurst 51, 56, /Lamontagne 32, 59, /Howard Rice 58, 64, 70, 110, /Stephen Robson 72, /Friedrich Strauss 29 bottom, 35, /Mel Watson 26, 50

John Glover 4, 11, 16, 31, 41, 44, 48, 86, 106, 107, 115

Octopus Publishing Group Limited 75, 80 bottom, 118 bottom, /Michael Boys 14, 33, 37, 42, 71, 93, 103 top, /Peter Burt 22 left, 22 right, 23 left, 23 right, /Jerry Harpur 61 top, 63, 76, 94, 103 bottom, 118 top left, 118 top right, /Andrew Lawson 83, 92, /Peter Myers 29 top, 40, 73, /Howard Rice 7, 17, 89, /Gareth Sambidge 34, /George Wright 12, 39, 61 bottom, 74, 80 top, 100, 102

Jerry Harpur/Dr. John Rivers, Balscote 8, /RHS Wisley 9

Andrew Lawson/Eastgrove Cottage, Worcs. 2-3, 10, 13, 15, 28, 38, 43, 57, 60, 62, 65, 66, 68, 84, 90, 104, 108, 112

Harry Smith Collection 24, 30, 105, 114, 116

Cover photography

Garden Picture Library/J.S. Sira front cover bottom
/Friedrich Strauss front cover top left

Octopus Publishing Group Limited/Michael Boys front cover top right

Editorial Manager: Jane Birch
Executive Art Editor: Peter Burt
Designer: Anthony Cohen
Picture Research: Christine Junemann
Production Controller: Ian Paton